# Advances in Marine Mechanical and Structural Engineering—2nd Edition

# Advances in Marine Mechanical and Structural Engineering—2nd Edition


Guest Editors

**Chenfeng Li**
**Kun Liu**
**Bin Liu**



MDPI

Basel • Beijing • Wuhan • Barcelona • Belgrade • Novi Sad • Cluj • Manchester

*Guest Editors*
Chenfeng Li
College of Shipbuilding Engineering
Harbin Engineering University
Harbin
China

Kun Liu
School of Naval Architecture and Ocean Engineering
Jiangsu University of Science and Technology
Zhenjiang
China

Bin Liu
Green & Smart River-Sea-Going Ship, Cruise and Yacht Research Centre
Wuhan University of Technology
Wuhan
China

*Editorial Office*
MDPI AG
Grosspeteranlage 5
4052 Basel, Switzerland

This is a reprint of the Special Issue, published open access by the journal *Journal of Marine Science and Engineering* (ISSN 2077-1312), freely accessible at: https://www.mdpi.com/journal/jmse/special_issues/O25DZO56E5.

For citation purposes, cite each article independently as indicated on the article page online and as indicated below:

Lastname, A.A.; Lastname, B.B. Article Title. *Journal Name* **Year**, *Volume Number*, Page Range.

**ISBN 978-3-7258-6536-9 (Hbk)**
**ISBN 978-3-7258-6537-6 (PDF)**
**https://doi.org/10.3390/books978-3-7258-6537-6**

# Contents

**Chenfeng Li, Kun Liu and Bin Liu**
Advances in Marine Mechanical and Structural Engineering—2nd Edition
Reprinted from: *J. Mar. Sci. Eng.* **2026**, *14*, 65, https://doi.org/10.3390/jmse14010065 . . . . . . **1**

**Lei Ao, Fuyou Li, Bin Liu, Nan Zhao and Junlin Deng**
Experimental and Numerical Analysis of the Collapse Behaviour of a Cracked Box Girder Under Bidirectional Cyclic Bending Moments
Reprinted from: *J. Mar. Sci. Eng.* **2025**, *13*, 1802, https://doi.org/10.3390/jmse13091802 . . . . . **4**

**Dong In Kim, Nak-Kyun Cho, Jin-Ha Hwang, Yu Yao Lin and Do Kyun Kim**
Engineering Critical Assessment of IMO Type C Tanks: A Comparative Study of Shell and Solid Element Models
Reprinted from: *J. Mar. Sci. Eng.* **2025**, *13*, 2185, https://doi.org/10.3390/jmse13112185 . . . . . **21**

**Jinsong Xia, Zhanyang Chen, Nan Zhao, Weidong Zhao, Qin Tang and Shijian Cai**
Free-Drop Experimental and Simulation Study on the Ultimate Bearing Capacity of Stiffened Plates with Different Stiffnesses under Slamming Loads
Reprinted from: *J. Mar. Sci. Eng.* **2024**, *12*, 1291, https://doi.org/10.3390/jmse12081291 . . . . . **45**

**Ye Lu, Liye Li and Qiu Jin**
A Comparative Study on the Calculation Methods of Nonlinear Springing of Large Containerships
Reprinted from: *J. Mar. Sci. Eng.* **2025**, *13*, 1226, https://doi.org/10.3390/jmse13071226 . . . . . **60**

**Yuxin Pan, Yuanyuan Wang, Fengyuan Liu and Gang Xu**
Effect of Damping Plate Parameters on Liquid Sloshing in Cylindrical Tanks of Offshore Launch Platforms
Reprinted from: *J. Mar. Sci. Eng.* **2025**, *13*, 1448, https://doi.org/10.3390/jmse13081448 . . . . . **74**

**Mengzhen Li, Xiaolong Liu, Mohammad Yazdi and Wei Chen**
Vibration Characteristic Analysis of Sandwich Composite Plate Reinforced by Functionally Graded Carbon Nanotube-Reinforced Composite on Winkler/Pasternak Foundation
Reprinted from: *J. Mar. Sci. Eng.* **2024**, *12*, 2157, https://doi.org/10.3390/jmse12122157 . . . . . **98**

**Chenfeng Li, Tenglong Jin, Zequan Chen and Guanchen Wei**
A De-Nesting Hybrid Reliability Analysis Method and Its Application in Marine Structure
Reprinted from: *J. Mar. Sci. Eng.* **2025**, *12*, 2221, https://doi.org/10.3390/jmse12122221 . . . . . **120**

**Guocai Chen, Xueliang Wang, Nan Zhao, Zhentao Jiang, Fei Li, Haozheng Chen, et al.**
Study on Strain Field Reconstruction Method of Long-Span Hull Box Girder Based on iFEM
Reprinted from: *J. Mar. Sci. Eng.* **2024**, *12*, 1482, https://doi.org/10.3390/jmse12091482 . . . . . **142**

**Ze Jiang, Pengyu Wei, Yuntong Du, Jiayi Peng and Qingbo Zeng**
A Virtual Assembly Technology for Virtual–Real Fusion Interaction of Ship Structure Based on Three-Level Collision Detection
Reprinted from: *J. Mar. Sci. Eng.* **2024**, *12*, 1910, https://doi.org/10.3390/jmse12111910 . . . . . **163**

Journal of
*Marine Science and Engineering*

MDPI

*Editorial*

# Advances in Marine Mechanical and Structural Engineering—2nd Edition

**Chenfeng Li [1], Kun Liu [2] and Bin Liu [3,*]**

[1] College of Shipbuilding Engineering, Harbin Engineering University, Harbin 150001, China; lichenfeng@hrbeu.edu.cn
[2] School of Naval Architecture and Ocean Engineering, Jiangsu University of Science and Technology, Zhenjiang 212100, China; kunliu@just.edu.cn
[3] Green & Smart River-Sea-Going Ship, Cruise and Yacht Research Centre, Wuhan University of Technology, Wuhan 430063, China
* Correspondence: liubin8502@whut.edu.cn

In the advanced design of novel structures used in marine, mechanical, and structural engineering, a pivotal challenge lies in accurately predicting their strength, amidst the integration of new materials and structures, within the context of extreme marine environments and potential accidents.

Advances in marine, mechanical, and structural engineering include mechanical analyses of advanced materials, such as alloys and composite materials, and strength analyses of novel structures such as sandwich structures in ship superstructures and special structures in underwater vehicles, in order to ensure that marine structures remain lightweight, safe, and economical throughout their lifetimes. Thus, the Special Issue, "Advances in Marine Mechanical and Structural Engineering", was prepared to collect works relating to advances in marine, mechanical, and structural engineering in general. To continue this work, the present Special Issue, "Advances in Marine Mechanical and Structural Engineering—2nd Edition", was developed to include more works related to this topic.

Ultimate strength evaluation constitutes a fundamental aspect of structural safety assessment in ship design, playing a pivotal role in ensuring structural integrity of in-service ships. For enhancing the collapse characteristics of damaged ships, Ao et al. [1] carried out an experimental and numerical analysis of the ultimate strength of a cracked box girder subjected to bidirectional cyclic bending moments.

With the increasing demand for liquefied $CO_2$ carriers to support carbon capture, utilization, and storage, conventional stress-based design approaches have limitations because they neglect imperfections resulting from fabrication and material. To assess these flaws, Kim et al. [2] conducted an engineering critical assessment for a liquefied $CO_2$ cargo tank to evaluate the effect of finite element configuration on structural integrity in the presence of potential flaws.

Slamming impact is a typical load on a ship bow. Differing from previous studies on free-drop tests, Xia et al. [3] focuses on the ultimate bearing capacity and failure mechanism of the ship's bow under slamming loads. The dynamic ultimate bearing capacity of stiffened plates with different stiffnesses under lateral slamming loads was studied.

The vibration of large containerships induced by waves and its resulting fatigue damage have been the focus of research in the field of marine engineering. Based on potential flow theory, Lu et al. [4] investigated the nonlinear wave-induced vibration response of large containerships, including the superposition of sum and difference frequencies, by considering the influence of second-order hydrodynamic forces.

With the growing demand for space launches, safer and more flexible offshore rocket launch technologies are being developed to overcome the limitations of land-based launches. To solve one of the technical problems, Pan et al. [5] employed CFD simulation to analyze liquid sloshing within a cylindrical tank, both with and without baffles.

Sandwich composite plates are widely used in many industrial applications including marine fields. Li et al. [6] presented numerical investigations into the free vibration properties of a sandwich composite plate with two fiber-reinforced plastic face sheets and a functionally graded carbon nanotube-reinforced composite core made of functionally graded carbon nanotube-reinforced composite resting on a Winkler/Pasternak elastic foundation.

To meet the safety and economical requirements of marine structures, uncertainty has been widely studied as a key factor in the design and manufacture of marine structures. Traditional methods for hybrid reliability analysis usually require a nested optimization framework, which will lead to too many calls to the limit state function and result in poor computational efficiency. In response to this problem, Li et al. [7] creatively proposed a de-nesting hybrid reliability analysis method showing greater effectiveness.

A structural health-monitoring system can be used to gather local or global strain data and perform precise structural health management, which is a crucial step in efficiently ensuring the safety of the structure. Chen et al. [8] carried out experimental research on the long-span hull box girder based on inverse finite element method technology to ensure the structural safety of the hull box girder.

With the rapid advancement of new-generation information technology, the virtual–real fusion interaction has increasingly become a crucial technique for structural analysis to determine the strength envelope of hulls. Jiang et al. [9] proposed a virtual assembly method for a structural virtual–real fusion test based on the oriented bounding box (OBB) algorithm, the Devillers and Guigue algorithm, and differential triangle facets algorithm.

In summary, the articles presented in this Special Issue cover broad research topics related to advances in marine mechanical and structural engineering, guiding readers through the best analysis approach.

**Author Contributions:** Writing—original draft preparation, B.L.; writing—review and editing, C.L. and K.L. All authors have read and agreed to the published version of the manuscript.

**Conflicts of Interest:** The authors declare no conflicts of interest.

## References


1. Ao, L.; Li, F.; Liu, B.; Zhao, N.; Deng, J. Experimental and Numerical Analysis of the Collapse Behaviour of a Cracked Box Girder Under Bidirectional Cyclic Bending Moments. *J. Mar. Sci. Eng.* **2025**, *13*, 1802. [CrossRef]
2. Kim, D.I.; Cho, N.-K.; Hwang, J.-H.; Lin, Y.Y.; Kim, D.K. Engineering Critical Assessment of IMO Type C Tanks: A Comparative Study of Shell and Solid Element Models. *J. Mar. Sci. Eng.* **2025**, *13*, 2185. [CrossRef]
3. Xia, J.; Chen, Z.; Zhao, N.; Zhao, W.; Tang, Q.; Cai, S. Free-Drop Experimental and Simulation Study on the Ultimate Bearing Capacity of Stiffened Plates with Different Stiffnesses under Slamming Loads. *J. Mar. Sci. Eng.* **2024**, *12*, 1291. [CrossRef]
4. Lu, Y.; Li, L.; Jin, Q. A Comparative Study on the Calculation Methods of Nonlinear Springing of Large Containerships. *J. Mar. Sci. Eng.* **2025**, *13*, 1226. [CrossRef]
5. Pan, Y.; Wang, Y.; Liu, F.; Xu, G. Effect of Damping Plate Parameters on Liquid Sloshing in Cylindrical Tanks of Offshore Launch Platforms. *J. Mar. Sci. Eng.* **2025**, *13*, 1448. [CrossRef]
6. Li, M.; Liu, X.; Yazdi, M.; Chen, W. Vibration Characteristic Analysis of Sandwich Composite Plate Reinforced by Functionally Graded Carbon Nanotube-Reinforced Composite on Winkler/Pasternak Foundation. *J. Mar. Sci. Eng.* **2024**, *12*, 2157. [CrossRef]
7. Li, C.; Jin, T.; Chen, Z.; Wei, G. A De-Nesting Hybrid Reliability Analysis Method and Its Application in Marine Structure. *J. Mar. Sci. Eng.* **2024**, *12*, 2221. [CrossRef]


 https://doi.org/10.3390/jmse14010065

8. Chen, G.; Wang, X.; Zhao, N.; Jiang, Z.; Li, F.; Chen, H.; Wei, P.; Zhang, T. Study on Strain Field Reconstruction Method of Long-Span Hull Box Girder Based on iFEM. *J. Mar. Sci. Eng.* **2024**, *12*, 1482. [CrossRef]
9. Jiang, Z.; Wei, P.; Du, Y.; Peng, J.; Zeng, Q. A Virtual Assembly Technology for Virtual–Real Fusion Interaction of Ship Structure Based on Three-Level Collision Detection. *J. Mar. Sci. Eng.* **2024**, *12*, 1910. [CrossRef]



 https://doi.org/10.3390/jmse14010065

Journal of
*Marine Science and Engineering*

MDPI

*Article*

# Experimental and Numerical Analysis of the Collapse Behaviour of a Cracked Box Girder Under Bidirectional Cyclic Bending Moments

**Lei Ao [1,2], Fuyou Li [2,3], Bin Liu [1,2,*], Nan Zhao [4] and Junlin Deng [5,*]**

1 Sanya Science and Education Innovation Park, Wuhan University of Technology, Sanya 572000, China; lei_ao@whut.edu.cn
2 Green & Smart River-Sea-Going Ship, Cruise and Yacht Research Center, Wuhan University of Technology, Wuhan 430063, China; lifuyou@whut.edu.cn
3 School of Naval Architecture, Ocean and Energy Power Engineering, Wuhan University of Technology, Wuhan 430063, China
4 China Ship Scientific Research Centre, Wuxi 214082, China; zhaonan702@126.com
5 School of Mechanical and Marine Engineering, Beibu Gulf University, Qinzhou 535011, China
* Correspondence: liubin8502@whut.edu.cn (B.L.); junlin.deng@163.com (J.D.)

**Abstract:** This study presents an integrated experimental and numerical investigation into the collapse characteristics of a cracked box girder subjected to bidirectional cyclic bending moments. An experimental test involving a box girder specimen with a prefabricated transverse crack on the deck panel is conducted under four-point bending to evaluate the influence of cracking on ultimate strength under cyclic loading. The findings are reported through load–displacement curves, strain measurements, and observations of both global and localised structural failure modes, demonstrating strong consistency with finite element simulations conducted using ABAQUS software (version 2022). The results reveal that cyclic loading prior to ultimate capacity induces negligible stiffness reduction in the box girder structure, consistent with the structural behaviour under monotonic loading. The initial failure mechanism is attributed to local buckling of the deck plate, subsequently followed by significant plastic deformation around the crack tips, ultimately leading to global collapse. Parametric studies are carried out to evaluate the influence of key variables on the girder's residual strength, such as crack length, cyclic load amplitude and pattern.

**Keywords:** ultimate strength; cyclic load; box girder; crack; experiment; finite element analysis

## 1. Introduction

Ultimate strength evaluation constitutes a fundamental aspect of structural safety assessment in ship design, playing a pivotal role in ensuring structural integrity of in-service ships. Since its incorporation into the IACS (International Association of Classification Societies) Common Structural Rules, assessing the ultimate capacity of hull girders has become a standardised requirement for verifying longitudinal strength over a ship's operational life. Extensive research over recent decades has significantly advanced the understanding of the ultimate strength of ship structures [1–6].

During recent years, the increasing adoption of high-strength steels and larger ship dimensions has resulted in reduced hull girder stiffness, presenting new challenges for ship structural safety. With the advancement in manufacturing high-strength steel that exhibits superior yield strength and toughness, its application in ship structures has enabled the use of thinner sections and reduced scantlings. While yield strength and toughness govern the initiation of plastic deformation and crack propagation, the dimensions of structural scantlings significantly influence buckling strength. Maintaining a design balance between buckling strength and material yielding necessitates further investigation into the ultimate strength of ship structures.

Structural failure incidents caused by insufficient longitudinal strength still occur (see, for example, Sumi et al. [7]). It indicates that the failure of hull girders in severe sea states is fundamentally a dynamic collapse process involving cyclic loading. Conventional assessment methods, which typically apply a monotonic bending moment under hogging or sagging conditions, fail to adequately capture structural behaviour under cyclic loading conditions. Notably, residual strength following repeated loading can be significantly lower compared to the instantaneous strength observed under the action of monotonic loading, exposing limitations in quasi-static analysis approaches.

Early studies primarily identified plastic accumulation as a failure mechanism, a phenomenon that has recently gained significant attention in ship hull structural analysis under cyclic loading conditions [8–10]. Li et al. [11] examined the ultimate strength of plate structures subjected to cyclic loading combined with lateral load, indicating that the number of cycles has significant effects on the plate's collapse modes. However, their analysis omitted material fracture behaviour, assuming the elastic–perfectly plastic material model. For better predicting residual strain accumulation under cyclic loading, more sophisticated material constitutive laws are required, incorporating a combined isotropic–kinematic hardening model accounting for the Bauschinger effect and cyclic ratcheting, especially for the higher amplitudes and multiple cycles. Chaboche et al. [12] pioneered the development of a mechanical constitutive material model that incorporates mixed hardening behaviour under cyclic loading conditions. Subsequent work by Li et al. [13] and Cui et al. [14] demonstrated that residual strength strongly correlates with the residual plastic strain at the unloading point in cyclic loading; i.e., higher residual strains lead to more pronounced strength reductions. When the applied load remains significantly below the ultimate capacity, the accumulated plastic deformation during cyclic loading is negligible, with each cycle's loading path nearly coinciding with the previous unloading path.

Under the action of longitudinal bending moments, the primary failure mode of a hull girder is typically nonlinear buckling on the compressive side. Recent studies have advanced the understanding of cyclic loading effects [15–17]. Sumi et al. [7] analysed a longitudinal strength failure incident of a container ship, revealing that the bottom plates could buckle at 90% of the ultimate hogging moment. Hu et al. [18] observed rapid box girder strength degradation under bidirectional cyclic loading, while Liu and Guedes Soares [19] demonstrated that cyclic loading can induce local fractures through plastic accumulation. Li et al. [20] introduced an incremental collapse method to evaluate cyclic ultimate structural capacity, finding that material hardening has a negligible impact on global bending response. However, a significant limitation of the existing literature is its predominant focus on numerical simulations, lacking experimental studies to validate these numerical models.

Few experimental investigations have employed simplified box girder specimens due to the complexity of large-scale testing. Deng et al. [21] analysed the ultimate strength of hull girders under unidirectional cyclic loading, in which the applied load reached the structural ultimate strength during the first cycle. The study primarily focused on the post-ultimate strength behaviour of the hull girders, and it did not consider bidirectional cyclic loading that more accurately represents the actual sagging and hogging experienced by ships in service. Song et al. [22] examined the response of a hull girder under bidirectional ultimate loading, pointing out that sustained exposure to ultimate condition leads to a progressive accumulation of plastic deformation, which reduces structural stiffness. The applied load also reached the ultimate strength within the first cycle, without accounting for the potential influence of cyclic loading below the ultimate strength level.

Nonlinear finite element methods can assist in estimating the structural ultimate capacity of structures, which have been widely applied to simulate various complex behaviours, including geometric and material nonlinearities. Chen et al. [23] utilised nonlinear finite element simulations for the welding process, investigating the influence of heat input variations on the performance of stiffened plates under cyclic loading. Li and Chen [24] analysed the maximum load-carrying capacity of stiffened panels subjected to simultaneous axial cyclic loading and lateral load, revealing a significant decrease in strength under the combined loading conditions. Cui et al. [25] carried out numerical simulations to evaluate how welding-induced deformation and residual stresses affect hull girders' strength under monotonic or cyclic bending moments.

Crack damage is a common and critical issue over the lifetime of ship hull structures in operation, significantly reducing the structural stiffness and load-bearing capacity. While monotonic loading effects on cracked structures have been extensively studied [26–29], recent attention has shifted to cyclic loading scenarios [30–32], revealing complex interactions between crack propagation and plastic accumulation. Cyclic loading can drive crack tip blunting or extension, and excessive load amplitude or a high number of cycles can lead to progressive crack propagation. Xia et al. [33] evaluated the residual strength of five cracked plates subjected to various loading scenarios, observing that the decrease in strength caused by cumulative plastic damage was less pronounced in relatively thicker plates. Hu et al. [34] experimentally analysed the strength degradation behaviour of cracked box girders under ultimate bending moments, and the results indicate that crack propagation accelerates the reduction in ultimate bending moment.

This study explores the residual strength of cracked box girders under cyclic loading, with particular emphasis on pre-ultimate unloading cycle effects. A cracked box girder specimen is designed for cyclic loading tests, with a prefabricated crack on the deck plate. The experimental outcomes, including the load–displacement behaviour, ultimate strength value, and failure modes, are contrasted with the findings from nonlinear finite element simulations. Additionally, the impact of various parameters such as crack length, cyclic load amplitude, and loading pattern on the ultimate loading capacity of cracked hull girders is analysed and discussed.

## 2. Experimental Details

The experimental programme evaluates the residual strength of a cracked hull girder specimen exposed to cyclic bending moments. A pure bending experiment is conducted for a benchmark study to analyse the ultimate strength behaviour of hull girders subjected to alternating cyclic bending loads.

### 2.1. Specimen Overview

The specimen's main dimensions are provided in Table 1, while the scantlings of its structural components are presented in Figure 1. The specimen comprises three segments: a central test segment flanked by two support segments on either end. The specimen is reinforced by an orthogonal angle-bar stiffened system at both ends of these segments (L-L for loading and S-S for supporting) to simulate the constraints of transverse bulkheads. A transverse crack (250 mm length × 10 mm width) is prefabricated on the deck plate of the test segment to simulate an initial imperfection in a ship deck structure. The crack length is set to half the specimen breadth and is located between two longitudinal stiffeners. Since these stiffeners provide crack-arresting capability, the crack is designed to terminate at both ends against them. This configuration references a mid-span crack scenario in the deck. The crack width was selected to facilitate fabrication.

**Table 1.** Geometric specifications of the test sample.

| Specimen | Unit | Dimension |
|---|---|---|
| Overall Length | mm | 2400 |
| Test section length | mm | 800 |
| Breadth | mm | 500 |
| Depth | mm | 300 |
| Longitudinal stiffener | mm | FB40 × 5.7 |
| Plate thickness of test segment | mm | 2.8 |
| Plate thickness of support segment | mm | 4.8 |
| Bulkhead | mm | L75 × 50 × 7.7 |

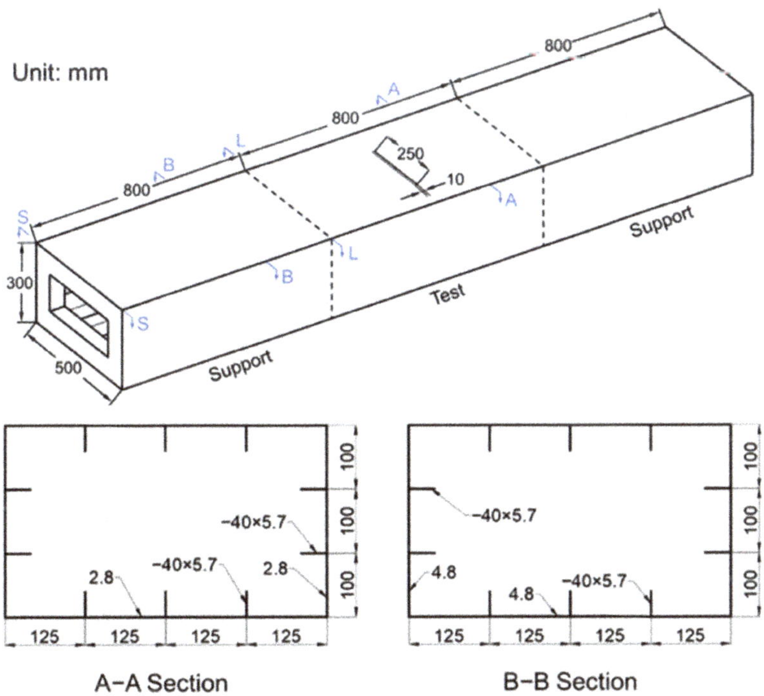


**Figure 1.** Geometric configuration and main section of the specimen.

The test segment of specimen is fabricated using low-carbon steels with two thicknesses of 2.8 mm and 5.7 mm. Quasi-static uniaxial tensile tests are conducted to characterise the mechanical properties of the steels. [35]. The results of the tensile tests for the two materials are presented in Figure 2, with yield stresses of 299 MPa and 324 MPa, respectively. The resulting engineering stress–strain curves are employed to analyse the material true stress–strain relationship before necking, which is then used for subsequent finite element simulations.

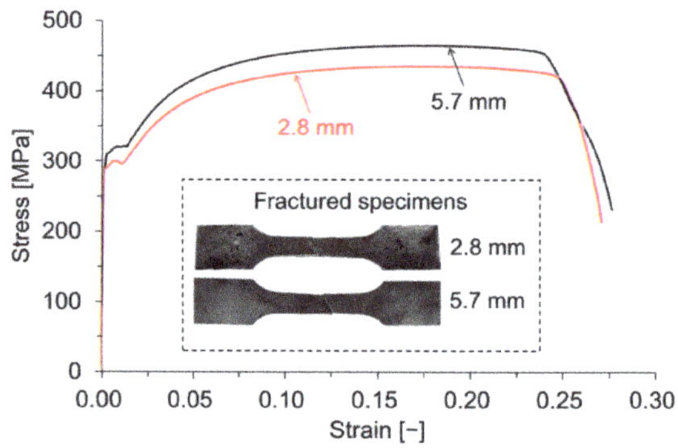


**Figure 2.** Engineering stress–strain relationship.

## 2.2. Experimental Setup

A four-point bending configuration is employed to induce vertical bending moments in the specimen, as shown in Figure 3. The specimen is mounted on two round steel bars resting on a rigid foundation (S-S section in Figure 1), in order to simulate the simply supported boundary conditions. Concentrated vertical loads are enforced by a hydraulic cylinder with a bolted loading beam, which transfers the forces to the box girder's cross frames through an additional set of round steel bars (L-L section in Figure 1).

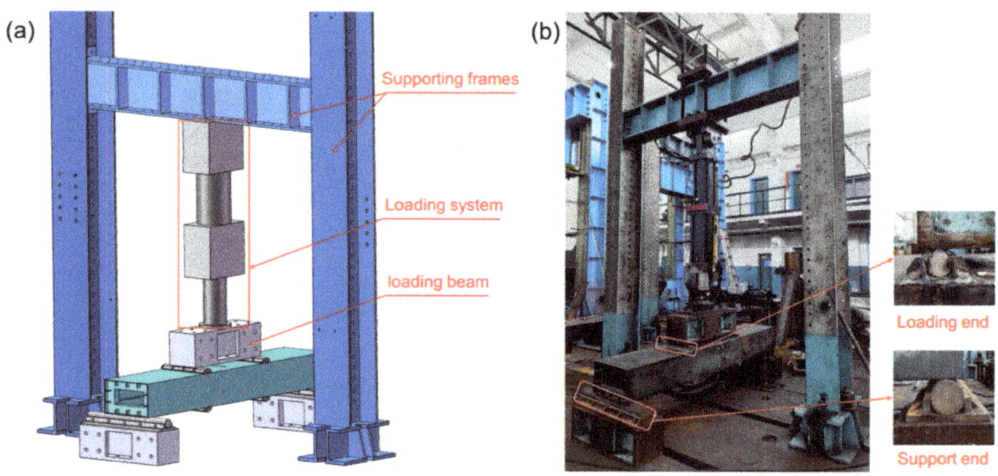


**Figure 3.** Experimental setup. (**a**) schematic diagram and (**b**) actual image.

The specimen is subjected to bidirectional cyclic loading over five successive loading cycles, with progressively increasing amplitude in each subsequent cycle, and finally is loaded until structural collapse occurs (see Figure 4). In the experiment, the specimen is inverted to achieve forward and reverse loadings, i.e., sagging and hogging bending moments. The initial five loading cycles are force-controlled at a rate of 0.5 kN/s until reaching the given value, while the final loading is applied via displacement control at a rate of 0.05 mm/s until structural collapse. Both force-controlled and displacement-controlled loading are regarded as quasi-static, and the output force is recorded by a load cell in the loading system.

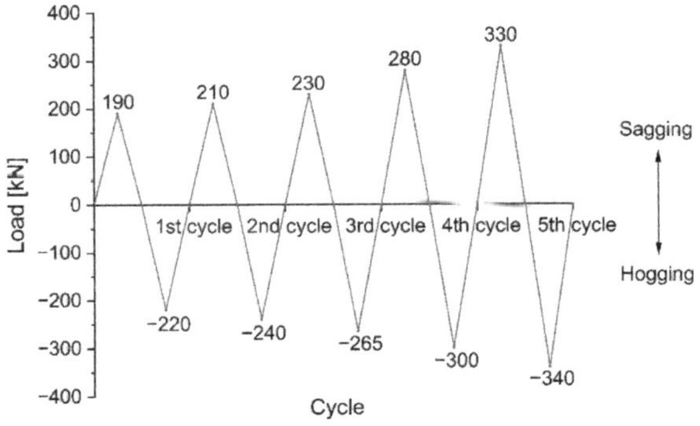


**Figure 4.** History of cyclic loading in the experiment.

Initial imperfections in the specimen, including initial deformations and welding-induced residual stresses, are inevitably introduced during fabrication. To mitigate their effects, preloading is applied to partially relieve the welding residual stresses, an approach consistent with methods reported in Refs. [5,6]. Consequently, the effect of welding residual stresses is excluded from both the experimental and numerical analyses.

Strain gauges oriented in three directions are installed in areas with expected stress concentrations, especially around the crack tips. Some critical measurement positions are depicted in Figure 5. A preliminary numerical analysis had been undertaken to optimise the gauge arrangement by searching the stress concentration regions.

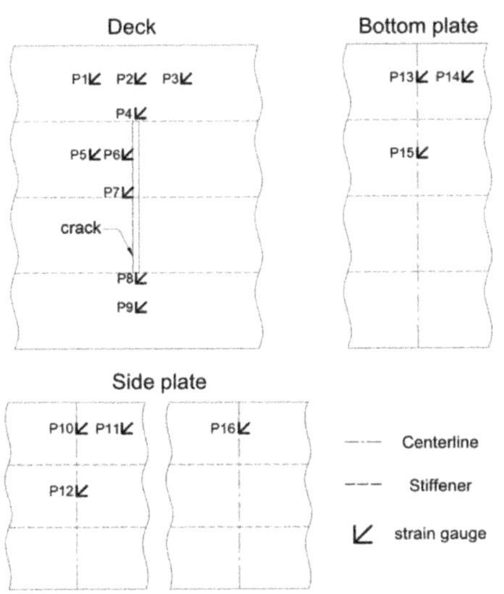


**Figure 5.** Layout of strain gauges.

Four displacement transducers are positioned at the bottom of the specimen to measure the vertical deflection, as shown in Figure 6. Measurement points D1, D3 and D4 are positioned at the transverse frames of specimen, and D2 is located at the centre of the bottom plate's span. The measured vertical displacements of D1 and D3 can be used to analyse the rotation angle of the test segment by dividing by the length of the support segment (800 mm). The comparison of D1 and D4 can be used to examine the lateral inclination of a specimen.

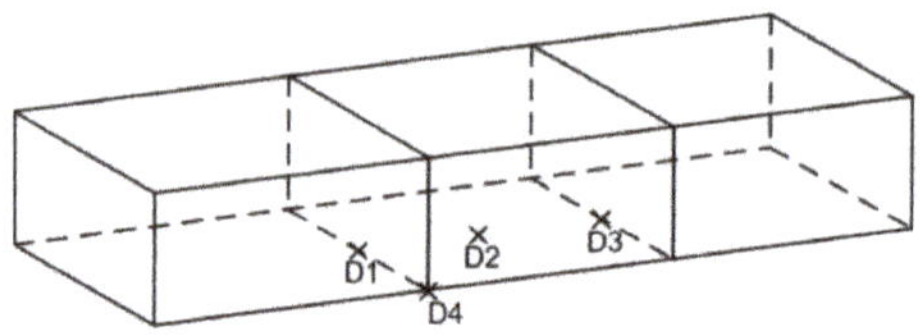


**Figure 6.** Location of displacement transducers.

The experimental procedure is summarised as follows.

(1) Preparation and preloading phase: Prior to the formal testing, the loading equipment and measurement instruments are calibrated. The specimen is subjected to three preloading cycles within the elastic range to partially release the residual welding stresses and to eliminate any potential interfacial gaps between specimen and test apparatus.
(2) First loading phase: The vertical loading is applied by the hydraulic cylinder in a force-controlled manner and the output force is recorded. After reaching the given load, the load is held for 10 s and then the unloading starts.
(3) Reverse loading and cyclic phase: The specimen is inverted for reverse loading and the displacement transducers are reinstalled. This procedure is repeated through five complete loading cycles.
(4) Ultimate loading phase: In the final monotonic loading stage, the specimen is subjected to displacement-controlled loading until structural collapse occurs. The collapse of a specimen is defined by a sudden large deformation accompanied by a noticeable load decrease. Afterwards, the loading process is immediately terminated, followed by a controlled unloading.

## 3. Finite Element Model

The nonlinear numerical simulation is conducted analysing the mechanical response of the box test specimen. The numerical model incorporates both geometric and material nonlinearities through an incremental–iterative solution approach based on the Newton–Raphson method. The specimen is modelled using S4R shell elements, which are four-node elements with reduced integration.

The applied loads and boundary constraints in the finite element model are presented in Figure 7. To simulate the simply supported boundary of the support ends, the nodal constraints are applied with Ux = Uy = Uz = 0 and Uy = Uz = 0, respectively. For each loading end, a reference point (RP1 or RP2) is defined at the centre of the loading line coupling with all points along the line. The five cyclic loading stages are implemented under force-controlled conditions to precisely simulate the experimental loading phases, and then the displacement-controlled loading is applied until the collapse of the structure.

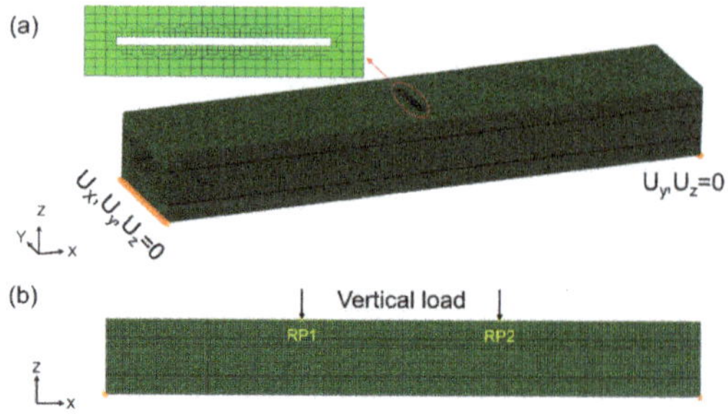


**Figure 7.** Finite element model of the box girder specimen. (**a**) three-view drawing and (**b**) side view.

The steel material properties are described through an elastic–plastic model that captures the actual stress–strain behaviour. Also, definitions are provided for Young's modulus, Poisson's ratio, and yield strength. The true stress ($\sigma_t$) and strain ($\varepsilon_t$) are derived from the engineering stress ($\sigma_e$) and strain ($\varepsilon_e$) using the following conversions:

$$\sigma_t = \sigma_e(1 + \varepsilon_e) \quad (1)$$

$$\varepsilon_t = \ln(1 + \varepsilon_e) \quad (2)$$

Initial geometrical imperfections in the specimen's plates and stiffeners are measured, and their overall deformations vary between −3.3 mm and 2.8 mm, as well as −6.9 mm and 7.3 mm, respectively. In the finite element model, imperfections are implemented using empirical formulae (Equations (3) and (4)), with amplitudes set to the measured mean values of 3.1 mm ($A_0$) and 7.1 mm ($B_0$) for plates and stiffeners, respectively.

$$w_{0P} = A_0 \sin\frac{m\pi x}{a} \sin\frac{\pi y}{b} \quad (3)$$

$$w_{0S} = B_0 \frac{z}{h_s} \sin\frac{\pi x}{a} \quad (4)$$

where $a$ and $b$ denote the length and width of the plate, respectively; $m$ is the longitudinal mode number of the plate; $h_s$ is the stiffener width.

The mesh convergence analysis results of the finite element model are presented in Figure 8. Three mesh sizes (5 mm, 10 mm and 20 mm) are evaluated, with the refined mesh near the crack region illustrated in Figure 7. The discrepancy in maximum bending load between the 5 mm and 10 mm meshes is only 0.84%, while that between the 10 mm and 20 mm meshes reaches 2.25%. Considering both computational efficiency and accuracy, the mesh size of 10 mm for the overall model and the refined mesh of 5 mm for the crack region are selected in subsequent finite element analyses. The mesh model contains a total of 49,484 elements as well as 49,435 nodes.

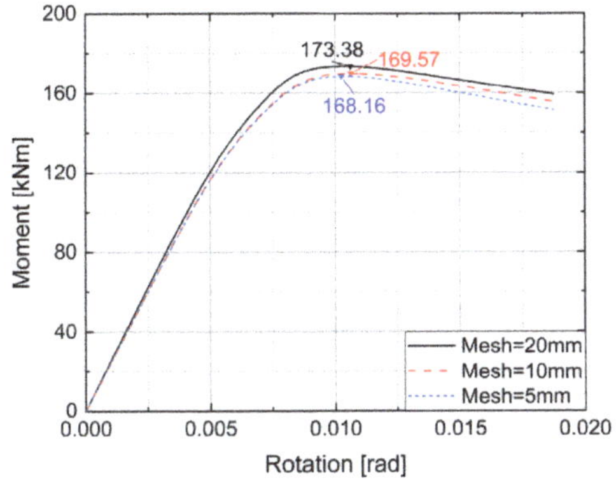


**Figure 8.** Comparison of bending moment–rotation curves with different mesh sizes.

In summary, the finite element model is developed with careful consideration of mesh selection, inclusion of geometric initial imperfections and the definition of the material constitutive model. These considerations ensure that the numerical simulation accurately captures the mechanical behaviour of cracked structures. The following section presents a comparison between the numerical and experimental results to demonstrate the accuracy of the finite element method.

## 4. Experiment and Numerical Results

### *4.1. Bending Moment vs. Rotation Curve*

#### 4.1.1. Experimental Result

The bending moment–rotation curves obtained from both experimental testing and numerical simulation for the box girder specimen are presented in Figure 9. The bending moment is obtained by multiplying the applied load at each loading line (half of the total load measured by the load cell) by the moment arm (800 mm). The resulting rotation angle is calculated by the vertical displacements of D1 and D3, presented in Figure 6, dividing by the length of the support segment (800 mm).

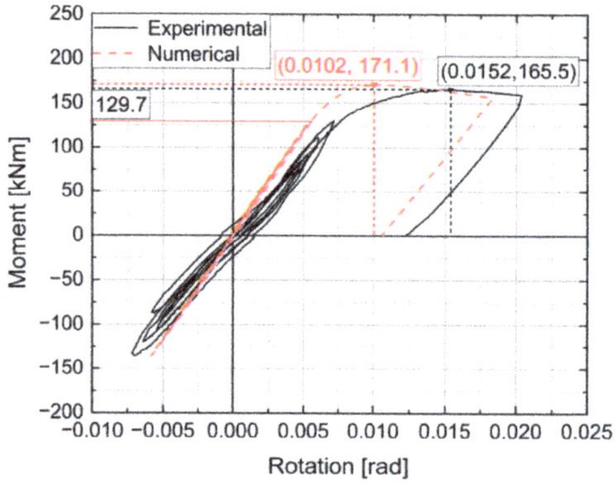


**Figure 9.** Experimental and simulated bending moment-rotation relationships.

The experimental bending load–rotation curve shows that its slope during the initial five loading cycles closely matches the linear phase of the final monotonic loading, indicating the negligible stiffness degradation due to cyclic loading. During the final monotonic loading phase, the load–displacement relationship remains linear up to a bending moment of 129.7 kNm, beyond which nonlinear behaviour emerges. When the load reaches 165.5 kNm, the specimen attains its ultimate bending capacity, signifying the onset of structural instability. Subsequent loading leads to a gradual decline in moment resistance.

#### 4.1.2. Numerical Result

The numerical ultimate load (171.1 kNm) agrees well with the experimental result (165.5 kNm), showing an error of 3.4%. This slight overestimation in the simulation may be attributed to welding residual stresses remaining in the test model after the preloading stage, which affects the ultimate bending moment. The numerical rotation angle corresponding to the ultimate load is notably smaller than the experimental one (0.0102 rad vs. 0.0152 rad).

Additionally, the elastic stiffness of the numerical curve diverges from the experimental data (see Figure 9). This discrepancy may be due to the experimental setup, where gaps between the round steel bars and the fixed supports cannot be completely eliminated through preloading. Similar phenomena have also been observed in published ultimate pure bending experiments of simple hull girders, as recorded by Xu et al. [36] and Hu et al. [34].

### 4.2. Collapse Mode

#### 4.2.1. Experimental Result

The experimental final deformation shape of the test segment under the sagging ultimate load is shown in Figure 10. During the initial five loading cycles, no significant deformation is detected near the deck crack or the side shell plate. Although stress concentration and plastic strain accumulation occur near the crack tip under cyclic loading, the resulting plastic strain remains relatively small and insufficient to induce crack propagation. This observation correlates with the bending moment–rotation curve, where structural plastic buckling typically manifests in the nonlinear stage.

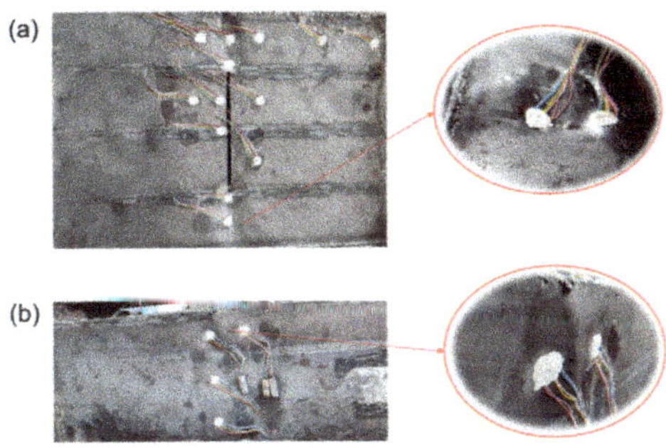


**Figure 10.** Experimental permanent deformation of (**a**) deck and (**b**) side plates.

As the prefabricated crack reduces the effective cross-sectional area and bending stiffness, the buckling initiates at both crack tips on the deck. Afterwards, the buckling deformation propagates laterally into the adjacent side shell regions, culminating in the global buckling of the specimen. Finally, severe buckling deformation is observed in the region near the crack located on the deck plates as well as the upper side ones.

#### 4.2.2. Numerical Result

The overall and local deformations of the test model obtained from the numerical analysis are displayed in Figure 11. Due to the crack's influence on the stress distribution across the deck, the structural failure of the model is initiated by localised buckling occurring at the crack tips. No significant deformation is observed on either side of the crack. This result aligns well with the experimental findings, confirming the accuracy and reliability of the numerical simulation.

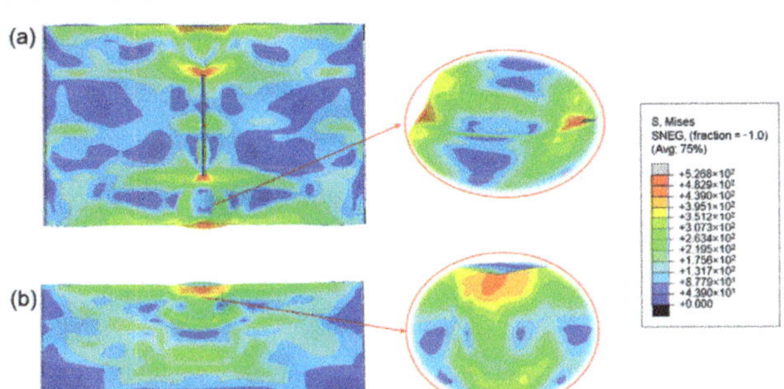


**Figure 11.** Numerical permanent deformation of (**a**) deck and (**b**) side plates.

Under the ultimate loading condition, the maximum plastic strain obtained at the crack tip element is 0.124. This level of plastic strain, 5 mm in size and 2.8 mm in thickness, is insufficient to initiate crack formation [37]. Furthermore, since no crack propagation was observed at the crack tip during the experiment, element removal was not incorporated in the finite element analysis.

### 4.3. Stress

#### 4.3.1. Experimental Result

The experimentally obtained equivalent stresses are derived from the data collected using triaxial strain gauges. The analysed stress values adopt the corresponding material yield stress (299 MPa and 324 MPa) when they exceed this threshold, since the stress calculation formulae in material mechanics are derived based on linear elastic assumptions. The stress values at the stress measurement points under sagging loads are defined as positive, with the corresponding hogging being negative, similar to the cyclic bending moment–rotation curve.

Six representative measurement points on the specimen are selected, and their stresses vs. bending moment curves are plotted in Figure 12. Points P6, P7, P8 and P9 are located around the deck crack, while P10 and P15 are positioned on the side and bottom plate, respectively (see Figure 5). In the first cycle loading stage (C1 in Figure 12), the concentrated stress at the crack front (P8) exceeds the material's yield stress (299 MPa) when the hogging bending moment reaches −88 kNm. In the stress release regions on the free edge of the crack, the stresses in P6 and P7 are much smaller than the one in P8 (Figure 12a,b). Due to the presence of a stiffener at the location of P7, part of the load is transferred to the deck, leading to the yielding of crack tips.

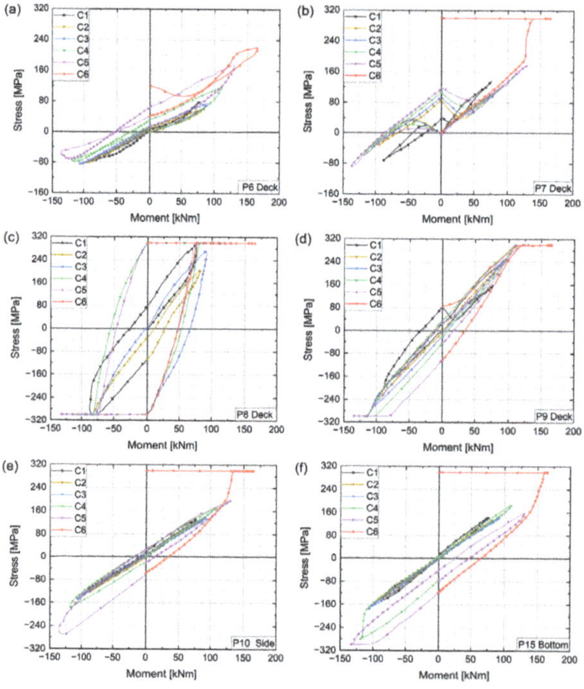


**Figure 12.** Experimental von Mises stresses at key measurement points on the specimen. (**a**) P6; (**b**) P7; (**c**) P8; (**d**) P9; (**e**) P10; (**f**) P15.

Residual stresses are accumulated at P9 and P15 due to the elastic instability occurring prior to the yield point during the cyclic loading process. This phenomenon is particularly obvious at P15 during the fourth hogging loading cycle (Figure 12f).

In the experiments, some measurement points are positioned symmetrically; for example, points P1 and P3 are longitudinally symmetrical, and points P10 and P16 are transversely symmetrical (see Figure 5). Their stress vs. bending moment curves are plotted in Figure 13, taking the fourth cyclic load as the typical example. Generally, points P1 and P3 exhibit nearly identical cyclic stress paths, indicating the highly symmetrical longitudinal deformation. Points P10 and P16 show some discrepancy during the loading, indicating the different deformations in the transverse direction. This may be attributed to the fact that the initial imperfections affect the structural buckling behaviour.

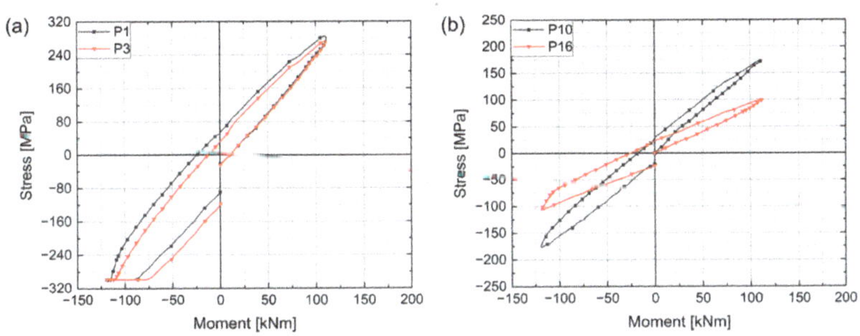


**Figure 13.** Experimental von Mises stresses of (**a**) P1 vs. P3 and (**b**) P10 vs. P16 during the 4th cycle.

4.3.2. Numerical Result

Figure 14 presents the numerical and experimental stress vs. bending moment curves at two representative measurement points of the structure (P9 on deck plate and P10 on side plate) during the fourth cyclic load. The overall trends of curves are effectively predicted by the numerical simulation. It is observed that the stresses obtained from numerical simulation are slightly lower than the experimental results. This discrepancy may be attributed to the fact that the welding residual stresses are not considered and the initial geometric imperfections of the specimen cannot be perfectly described by the finite element model. Numerous uncertainties inherent in actual structures necessitate certain simplifications in finite element simulations, which inevitably introduce minor discrepancies in the comparison. The stresses exhibit an approximately linear relationship with the applied bending moment during the loading phase, and the residual stresses emerge after unloading as a result of structural buckling.

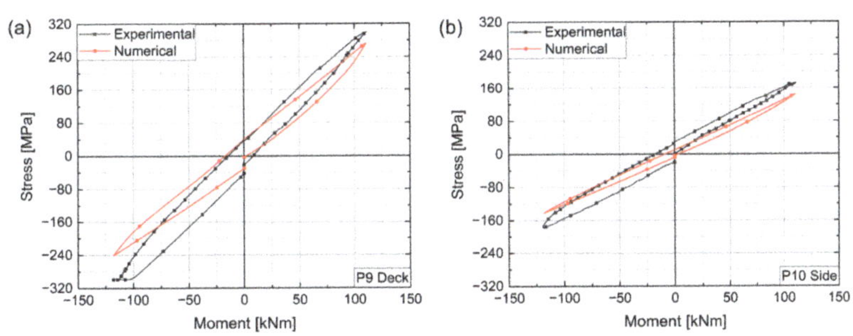


**Figure 14.** Comparison of experimental and numerical von Mises stresses of (**a**) P9 and (**b**) P10 during the 4th cycle.

## 5. Discussion

The following subsections explore various factors influencing the cracked hull girders' residual strength, including (1) crack, (2) cyclic load amplitude and pattern, and (3) material model selection.

### 5.1. Effect of Crack

Numerical comparisons of the bending moment–rotation curves under monotonic and multi-cyclic loadings are given in Figure 15. Generally, the curve in multiple cycles closely follows the path of the curve under monotonic loading, which may be due to the fact that residual plastic strains in multiple cycles slightly affect the structural ultimate strength. Since the applied load level stays beneath the girder's ultimate loading capacity in limited cycle numbers, the accumulated plastic deformation during cyclic loading is negligible, with each cycle's loading path nearly coinciding with the previous unloading path.

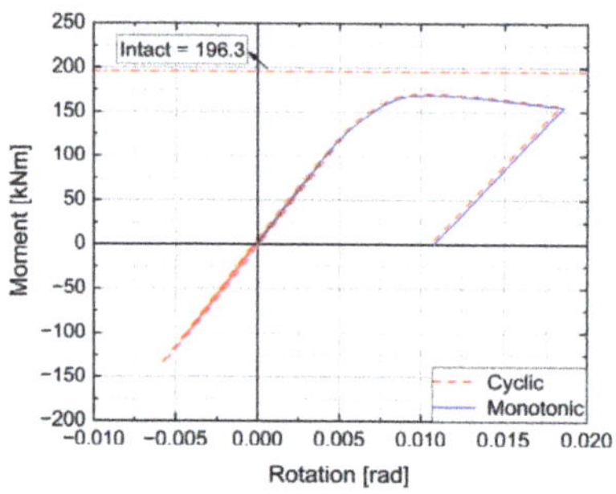


**Figure 15.** Numerical comparison of the bending load–rotation relationships subjected to monotonic and multi-cyclic loading.

To investigate the crack's effect on the girder's collapse behaviour, one more simulation is performed without crack damage, and the comparative results are shown in Figure 15. The intact model exhibits a maximum flexural moment of 196.3 kNm, while the presence of the crack leads to a notable reduction (12.8%) in sagging conditions.

To systematically investigate the crack length effects, the dimensionless crack length ratio is introduced, where $c$ represents crack length and $b$ denotes specimen width (500 mm). Here, three specific ratios ($c/b$ = 0.25, 0.5 and 0.75) are selected, and the resulting bending moment–rotation curves are shown in Figure 16. It reveals that the reduced initial stiffness of cracked specimens strongly affects the structural ultimate strength.

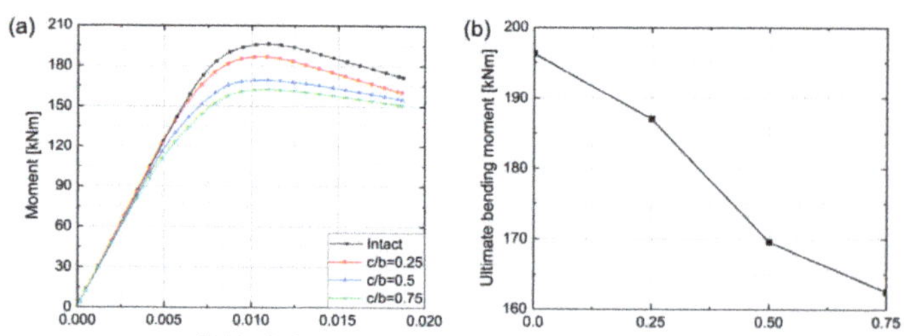


**Figure 16.** Influence of crack length under monotonic bending. (**a**) Bending moment vs. rotation curves and (**b**) Ultimate bending moment vs. crack length curve.

The ultimate strength exhibits a nearly linear reduction as the length of the crack grows, with a maximum decrease of 17.2% observed at $c/b = 0.75$. Notably, the rotation angle at ultimate loading remains relatively constant across all crack lengths. This indicates that crack growth primarily affects structural strength, while its effect on deformation is minimal.

### 5.2. Effect of Cyclic Load

#### 5.2.1. Load Amplitude

In the previous numerical simulations, the maximum cyclic bending load and the final ultimate flexural moment are 136.0 kNm and 171.1 kNm, respectively. To analyse the effect of cyclic load amplitude, two larger cyclic bending moments (148 kNm and 164 kNm) are selected. In the loading process, five bidirectional cyclic loadings and a final monotonic ultimate loading are carried out.

As illustrated in Figure 17, the residual angular displacement at zero load increases with the magnitude of cyclic loading due to the accumulation of inelastic deformation. It also means that the unloading path deviates slightly from the initial loading path in the case of a cyclic bending moment of 164 kNm. However, these phenomena slightly affect the girder's residual strength, as the failure mode is primarily governed by buckling behaviour without crack propagation.

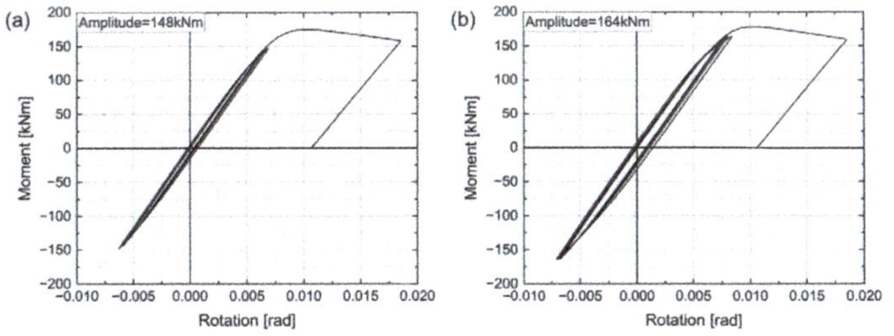


**Figure 17.** Bending moment–rotation curves under cyclic loading with amplitudes of (**a**) 148 kNm and (**b**) 164 kNm.

#### 5.2.2. Unidirectional Cyclic Loading

The effect of unidirectional cyclic loading on the hull structure's loading capacity was examined experimentally and numerically by Paik et al. [38], Deng et al. [21] and Song et al. [39]. Here, a numerical investigation into the ultimate strength of the test model subjected to unidirectional cyclic loading is carried out, compared to the effect of bidirectional cyclic loading (Figure 18). It is observed that the loading and unloading curves exhibit nearly identical paths, since the applied cyclic loading does not cause significant buckling deformation of the specimen.

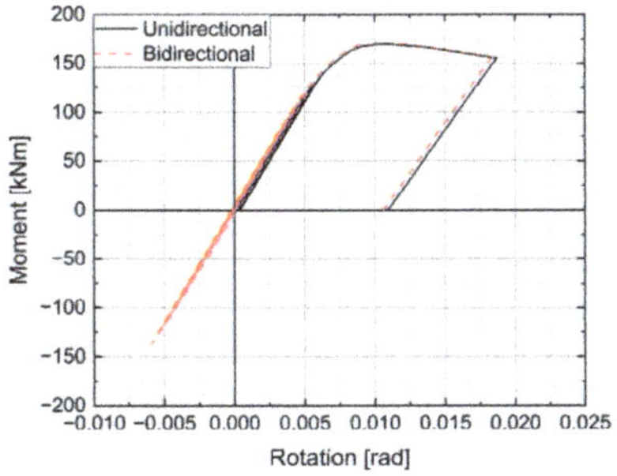


**Figure 18.** Bending moment–rotation curves under unidirectional and bidirectional cyclic loadings.

### 5.3. Effect of Material Model

The elastic–perfectly plastic material model is commonly used in the evaluation of thin plate structures' failure strength. To analyse the effect of simplification of the material model, one more simulation is performed in comparison with the definition of the true stress–strain relationship of the material. In the simplified elastic–plastic material model, the stress keeps constant after reaching the material yield point (299 MPa and 324 MPa). Their comparison using the flexural load–rotation curves is presented in Figure 19.

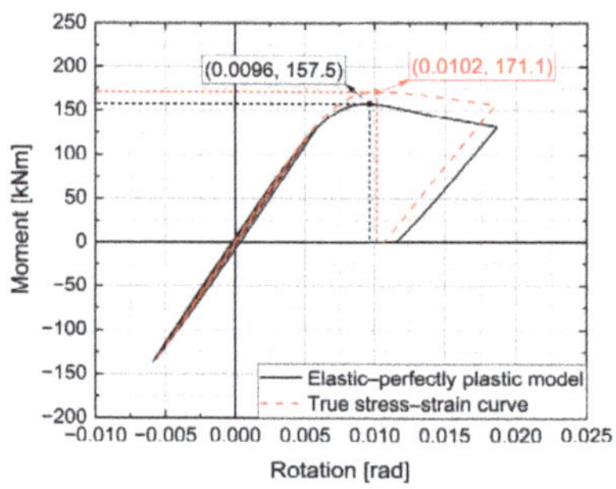


**Figure 19.** Comparison of material model selection using elastic–perfectly plastic model and true stress–strain curve.

Compared to the implementation of a true stress–strain curve, the use of the elastic–perfectly plastic model means that the nonlinear stage is reached earlier with a smaller rotation angle at ultimate loading. The resulting ultimate bending capacity for the elastic–perfectly plastic model is 157.5 kNm, with a 7.9% reduction compared to the case of the true stress–strain curve (171.1 kNm). This discrepancy is mainly due to the elastic–perfectly plastic model ignoring post-yield hardening behaviour of the material, which ultimately results in an underestimation of the structural ultimate strength.

## 6. Conclusions

The present research has addressed combined experimental and numerical investigations into the ultimate strength of a cracked box girder under bidirectional cyclic bending moments, illustrating the gradual collapse behaviour of the damaged structure.

The reduction in the ultimate strength of a structure due to cracking can be attributed to two primary factors: crack propagation at the tip leading to structural fracture failure and reductions in bending stiffness causing premature buckling failure. In the present experiment, the collapse of the hull girder was primarily governed by buckling phenomena without crack propagation. Local buckling occurs firstly at both ends of the crack and then gradually propagates to side plates, leading to global structural collapse. For real ship structures, the dominant failure mechanism, whether fracture or buckling, depends on the specific structural characteristics and requires further investigation.

At load levels significantly below the maximum load-carrying capacity of the structure, the accumulated plastic deformation during cyclic loading is negligible. The variation in the load amplitude and loading mode has no noticeable influence on the girder's ultimate strength within a constrained number of cycles. The application of the elastic–perfectly plastic material model causes a significant reduction in the structural ultimate strength based on numerical analysis.

In the present experiment, both the amplitude and number of cyclic loadings were limited. Future research could apply the loads closer to the structural ultimate strength under an increased number of cycles. Furthermore, since the welding residual stresses were not accurately quantified, their uncertainty introduced limitations in the result analysis. Future research could incorporate advanced measurement techniques to obtain precise data on these stresses.

**Author Contributions:** Conceptualization, L.A., F.L. and B.L.; methodology, L.A. and F.L.; software, F.L.; validation, L.A.; writing—original draft preparation, L.A. and F.L.; writing—review and editing, B.L., N.Z. and J.D.; supervision, L.A. and B.L. All authors have read and agreed to the published version of the manuscript.

**Funding:** This research was funded by the National Natural Science Foundation of China (Grant No. 52271326; 52301385; 52161042) and the Natural Science Foundation of Guangxi Province (Grant No. 2025GXNSFAA069191).

**Institutional Review Board Statement:** Not applicable.

**Informed Consent Statement:** Not applicable.

**Data Availability Statement:** Data presented in this article are available upon request from the corresponding author.

**Conflicts of Interest:** The authors declare no conflicts of interest.

## References

1. Saad-Eldeen, S.; Garbatov, Y.; Guedes Soares, C. Effect of corrosion severity on the ultimate strength of a steel box girder. *Eng. Struct.* **2013**, *49*, 560–571. [CrossRef]
2. Garbatov, Y.; Saad-Eldeen, S.; Guedes Soares, C. Hull girder ultimate strength assessment based on experimental results and the dimensional theory. *Eng. Struct.* **2015**, *100*, 742–750. [CrossRef]
3. Kim, D.K.; Lim, H.L.; Yu, S.Y. A technical review on ultimate strength prediction of stiffened panels in axial compression. *Ocean Eng.* **2018**, *170*, 392–406. [CrossRef]
4. Li, S.; Kim, D.K.; Benson, S. The influence of residual stress on the ultimate strength of longitudinally compressed stiffened panels. *Ocean Eng.* **2021**, *231*, 108839. [CrossRef]
5. Liu, B.; Gao, L.; Ao, L.; Wu, W. Experimental and numerical analysis of ultimate compressive strength of stiffened panel with openings. *Ocean Eng.* **2021**, *220*, 108453. [CrossRef]
6. Liu, B.; Yao, X.; Lin, Y.; Wu, W.; Guedes Soares, C. Experimental and numerical analysis of ultimate compressive strength of long-span stiffened panels. *Ocean Eng.* **2021**, *220*, 109633. [CrossRef]
7. Sumi, Y.; Fujikubo, M.; Fujita, H.; Kawagoe, Y.; Kidogawa, M.; Kobayashi, K.; Nakano, T.; Iwano, J.; Takahira, T.; Tamura, K. *Final Report of Committee on Large Container Ship Safety*; Committee on Large Container Ship Safety: Tokyo, Japan, 2015.
8. Fukumoto, Y.; Kusama, H. Local instability tests of plate elements under cyclic uniaxial loading. *J. Struct. Eng.* **1985**, *111*, 1051–1067. [CrossRef]
9. Cui, H.; Ding, Q. Ultimate strength and fracture failure of hull stiffened plates based on plastic accumulation under cyclic loading. *Ocean Eng.* **2022**, *261*, 112016. [CrossRef]
10. Chen, X.; Yue, J.; Song, Z.; Xu, R.; Rong, M. Study on ultimate strength for stiffened plates with elasto-plastic damages. *Ocean Eng.* **2023**, *279*, 114531. [CrossRef]
11. Li, D.; Chen, Z.; Chen, X. Numerical investigation on the ultimate strength behaviour and assessment of continuous hull plate under combined biaxial cyclic loads and lateral pressure. *Mar. Struct.* **2023**, *89*, 103408. [CrossRef]
12. Chaboche, J.L.; Van, K.D.; Cordier, G. Modelization of the strain memory effect on the cyclic hardening of 316 stainless steel. In Proceedings of the Transactions of the International Conference on Structural Mechanics in Reactor Technology, Berlin, Germany, 13–17 August 1979.
13. Li, S.; Hu, Z.; Benson, S. An analytical method to predict the buckling and collapse behaviour of plates and stiffened panels under cyclic load. *Eng. Struct.* **2019**, *199*, 109627. [CrossRef]
14. Cui, H.; Chen, Z.; Hu, R.; Ding, Q. Ultimate strength assessment of hull girders considering elastic shakedown based on Smith's method. *Ocean Eng.* **2024**, *293*, 116695.
15. Iijima, K.; Fujikubo, M. Cumulative collapse of a ship hull girder under a series of extreme wave loads. *J. Mar. Sci. Technol.* **2015**, *20*, 530–541. [CrossRef]
16. Zhang, X.; Paik, J.K.; Jones, N. A new method for assessing the shakedown limit state associated with the breakage of a ship's hull girder. *Ships Offshore Struct.* **2016**, *11*, 92–104.

17. Cui, H.; Yang, P. Ultimate strength assessment of hull girder under cyclic bending based on smith's method. *J. Ship Res.* **2018**, *62*, 77–88. [CrossRef]
18. Hu, K.; Yang, P.; Xia, T.; Song, Y.; Chen, B. Numerical investigation on the residual ultimate strength of cracked stiffened plates under extreme cyclic loads. *Ocean Eng.* **2022**, *244*, 110426. [CrossRef]
19. Liu, B.; Guedes Soares, C. Ultimate strength assessment of ship hull structures subjected to cyclic bending moments. *Ocean Eng.* **2020**, *215*, 107685. [CrossRef]
20. Li, S.; Hu, Z.; Benson, S. Progressive collapse analysis of ship hull girders subjected to extreme cyclic bending. *Mar. Struct.* **2020**, *73*, 102803. [CrossRef]
21. Deng, H.; Yuan, T.; Gan, J.; Liu, B.; Wu, W. Experimental and numerical investigations on the collapse behaviour of box type hull girder subjected to cyclic ultimate bending moment. *Thin Wall. Struct.* **2022**, *175*, 109204. [CrossRef]
22. Song, S.; Ehlers, S.; Von Bock Und Polach, F.; Braun, M. Ultra-low cycle fatigue of ship hull structure—An alternately-cyclically loaded four-point bending test of a large box girder. *Mar. Struct.* **2025**, *100*, 103732. [CrossRef]
23. Chen, Z.; Li, D.; Li, J.; Yi, J. The influence of welding-induced 3D residual stress and distortion on the ultimate strength of multi-stiffened thin plate structures under cyclic load. *Ocean Eng.* **2024**, *311*, 118797. [CrossRef]
24. Li, D.; Chen, Z. Progressive collapse response and ultimate strength evaluation of stiffened plates with welding residual stress under combined biaxial cyclic loads and lateral pressure. *Mar. Struct.* **2025**, *99*, 103703. [CrossRef]
25. Cui, H.; Hu, R.; Chen, Z.; Zheng, C. Research on ultimate strength of hull girder considering initial imperfections under monotonic/cyclic bending moments—A bulk carrier case. *Ocean Eng.* **2024**, *311*, 118862. [CrossRef]
26. Xu, M.C.; Garbatov, Y.; Guedes Soares, C. Residual ultimate strength assessment of stiffened panels with locked cracks. *Thin Wall. Struct.* **2014**, *85*, 398–410. [CrossRef]
27. Ao, L.; Wang, D.Y. Ultimate torsional strength of cracked stiffened box girders with a large deck opening. *Int. J. Nav. Archit. Ocean Eng.* **2016**, *8*, 360–374. [CrossRef]
28. Ao, L.; Wu, H.; Wang, D.Y.; Wu, W.G. Evaluation on the residual ultimate strength of stiffened plates with central dent under longitudinal thrust. *Ocean Eng.* **2020**, *202*, 107167. [CrossRef]
29. Shi, X.H.; Zhang, J.; Guedes Soares, C. Numerical assessment of experiments on the residual ultimate strength of stiffened plates with a crack. *Ocean Eng.* **2019**, *171*, 443–457. [CrossRef]
30. Feng, G.; Wang, Y.; Garbatov, Y.; Ren, H.; Guedes Soares, C. Experimental and numerical analysis of crack growth in stiffened panels. *Ships Offshore Struct.* **2021**, *16*, 980992. [CrossRef]
31. Xia, T.; Yang, P.; Song, Y.; Hu, K.; Qian, Y.; Feng, F. Ultimate strength and post ultimate strength behaviors of hull plates under extreme longitudinal cyclic load. *Ocean Eng.* **2019**, *193*, 106589. [CrossRef]
32. Hu, K.; Yang, P.; Xia, T.; Qin, D. Ultimate strength assessment of cracked stiffened box girders subjected to extreme cyclic bending moments. *Ocean Eng.* **2022**, *256*, 111496. [CrossRef]
33. Xia, T.; Yang, P.; Cui, H.; Song, Y. Experimental research on ultimate strength behaviors of cracked plates under uniaxial cyclic load. *Ocean Eng.* **2025**, *322*, 120514. [CrossRef]
34. Hu, K.; Zhang, F.; Xu, R.; Cui, H.; Yang, P. Experimental study on the ultimate strength of box-type hull girders subjected to extreme cyclic bending moments. *Ocean Eng.* **2025**, *339*, 122044. [CrossRef]
35. *GB/T 228.1*; Metallic Materials-Tensile Testing, Part I: Method of Test at Room Temperature. Standards Press of China: Beijing, China, 2021. (In Chinese)
36. Xu, S.; Liu, B.; Garbatov, Y.; Wu, W.; Guedes Soares, C. Experimental and numerical analysis of ultimate strength of inland catamaran subjected to vertical bending moment. *Ocean Eng.* **2019**, *188*, 106320. [CrossRef]
37. Liu, B.; Villavicencio, R.; Zhang, S.; Guedes Soares, C. A simple criterion to evaluate the rupture of materials in ship collision simulations. *Mar. Struct.* **2017**, *54*, 92–111. [CrossRef]
38. Paik, J.K.; Lee, D.H.; Noh, S.H.; Park, D.K.; Ringsberg, J.W. Full-scale collapse testing of a steel stiffened plate structure under cyclic axial-compressive loading. *Structures* **2020**, *26*, 996–1009. [CrossRef]
39. Song, Z.J.; Liu, Q.C.; Xu, R.J.; Yang, X.; Yue, J.X. Ultimate strength attenuation behaviours of stiffened panels under cyclic extreme loads based on test and numerical simulations. *Ocean Eng.* **2025**, *321*, 120447. [CrossRef]

Journal of
Marine Science
and Engineering

MDPI

*Article*

# Engineering Critical Assessment of IMO Type C Tanks: A Comparative Study of Shell and Solid Element Models

**Dong In Kim [1], Nak-Kyun Cho [2], Jin-Ha Hwang [3], Yu Yao Lin [1] and Do Kyun Kim [1,4,*]**

1 Department of Naval Architecture and Ocean Engineering, Seoul National University, Seoul 08826, Republic of Korea
2 Department of Manufacturing System and Design Engineering, Seoul National University of Science and Technology, Seoul 01811, Republic of Korea
3 Department of Mechanical Engineering, Pukyong National University, Busan 48513, Republic of Korea
4 Research Institute of Marine Systems Engineering, Department of Naval Architecture and Ocean Engineering, Seoul National University, Seoul 08826, Republic of Korea
* Correspondence: do.kim@snu.ac.kr

**Abstract:** In the present study, an Engineering Critical Assessment (ECA) is conducted for an International Maritime Organisation (IMO) Type C liquefied $CO_2$ ($LCO_2$) cargo tank to evaluate the effect of finite element configuration on structural integrity in the presence of potential flaws. With the increasing demand for $LCO_2$ carriers to support carbon capture, utilisation, and storage (CCUS), conventional stress-based design approaches outlined in the International Gas Carrier (IGC) Code have limitations because they neglect imperfections resulting from fabrication and material. To assess these flaws, the fracture mechanics-based ECA methodology, as prescribed by the BS 7910 standard, is applied to a bilobe IMO type C tank designed for cryogenic and pressurised conditions. The assessment integrates fracture toughness, stress intensity factor, and applied loads. Both the two-dimensional shell element model and the three-dimensional solid element model are developed and compared in terms of stress distribution, safety factor for fracture, and fatigue crack growth predictions. Results show that while shell models offer computational efficiency, solid models capture bending stresses and stress concentrations at geometric discontinuities more accurately, resulting in higher reliability in ECA outcomes. The comparative analysis highlights that the web and butt weld near the centre bulkhead are the most vulnerable regions, and fatigue crack growth is highly sensitive to input data, such as stress intensity factor range and fatigue crack growth laws. These findings provide practical guidance for applying ECA in bilobe $LCO_2$ tank design and safety assessment.

**Keywords:** fracture mechanics; Liquified $CO_2$; ammonia; Failure Assessment Diagram (FAD); stress intensity factor; stress distribution

## 1. Introduction

In response to global warming and climate change, reducing carbon emissions has become a critical priority across multiple sectors. Maritime transport accounts for approximately 3% of total global carbon dioxide ($CO_2$) emissions [1], and the annual emissions increased from 889 to 974 million tonnes between 2019 and 2024, corresponding to an average growth rate of 1.8% [2]. To address this challenge, the International Maritime Organization (IMO) has adopted a strategy to achieve net-zero greenhouse gas (GHG) emissions from international shipping by 2050. As intermediate goals, the IMO set indicative checkpoints of at least 20% (striving for 30%) reduction by 2030 and at least 40% (striving for 80%) reduction by 2040 [3]. In line with these goals, Liquefied Natural Gas (LNG) carriers and LNG-fuelled vessels have been introduced as transitional solutions, while the use of hydrogen-based fuels such as ammonia and methanol is expected to expand in the future [4]. At the same time, marine transportation also plays an important role in Carbon Capture, Utilisation and Storage (CCUS), as liquefied $CO_2$ should be delivered to offshore storage sites [5]. Consequently, the demand for liquefied $CO_2$ carriers, as well as liquefied gas tanks for alternative fuels, is estimated to increase. The development of advanced liquefied gas tank technologies is therefore essential to support the transition toward net-zero GHG emissions in shipping.

The IMO adopted the International Code for the Construction and Equipment of Ships Carrying Liquefied Gases in Bulk (IGC Code) to ensure the safe transport of liquefied gases [6]. The IGC Code prescribes structural integrity assessment procedures for liquefied gas tanks, primarily based on allowable stress design, in which applied stresses are compared with stress limits derived from material resistance and safety factors. However, this approach assumes intact structures without imperfections, leaving no provision for evaluating potential flaws. With the trend toward larger liquefied gas tanks to improve transport efficiency, thicker plates and increased principal dimensions have introduced greater challenges in fabrication and inspection. For instance, post-weld heat treatment (PWHT), which is used to relieve residual stresses, is constrained by the physical limitations of furnace capacity. Similarly, non-destructive testing (NDT) becomes more complex in large structures due to difficulties in selecting test regions, while thicker plates increase uncertainty in inspection results. Consequently, there is a growing need to incorporate potential flaws into structural integrity assessments. To address this, fracture mechanics–based assessment procedures have been introduced as a complement to the traditional stress-based approach.

As a fracture mechanics-based assessment method, Engineering Critical Assessment (ECA) integrates fracture mechanics parameters that indicate crack severity with applied stress to evaluate the acceptability of potential defects. This approach considers material resistance, crack geometry, and loading conditions to provide a quantitative prediction of failure risk and to support design and maintenance strategies aimed at ensuring structural reliability. The ECA is increasingly applied to pressure vessels, ship structures, cryogenic tanks, and other systems where quantifying the safety margin of flaws under fatigue and ultimate loads is essential. Liquefied $CO_2$ cargo tanks represent a typical example of such applications. For $CO_2$ carriers, carbon dioxide must be transported in liquefied form to achieve sufficient storage efficiency. Since the triple point of $CO_2$ lies above atmospheric pressure, pressurised vessels are required. In the IGC code, pressurised vessels requiring a design vapour pressure above 2 atm are classified as IMO type C independent tanks. Typically, small-scale $CO_2$ ship cargos operate at −35 °C and 19 atm, while large-scale cargos operate at −55 °C and 8 atm. Both types demand higher design pressures compared with other marine cargo tanks, resulting in the use of thicker plates. For large cargo tanks, plate thickness can reach up to 50 mm. However, the IGC Code only specifies structural requirements up to 40 mm; therefore, the International Association of Classification Societies (IACS) has extended the safety standard to cover plate thicknesses up to 50 mm.

The requirements for these materials are specified in the UR W1 standard [7]. According to this standard, IMO type C independent tanks generally require post-weld stress relief heat treatment. As an alternative, this procedure may be waived if an ECA is performed and approved by either the IACS or recognised international standards. These

ECA standards include those published by the British Standards Institution (BS 7910), EDF Energy (R6), and the American Petroleum Institute (API 579) [8–10].

Since the classification standard does not prescribe detailed analysis conditions, an appropriate and representative ECA setup must be carefully defined. In particular, the stress intensity factor (SIF), which quantifies the severity of a crack, is highly sensitive to the crack geometry and imperfections. Moreover, accurate evaluation requires consideration of stress distribution through the plate thickness, as the stress singularity at the crack tip is influenced by local variations. To reflect the different stress states of various structures in SIF calculations, ECA standards employ stress linearisation for membrane and bending components. These sectional stress distributions are typically obtained through numerical analysis, such as the finite element method (FEM). Although ship structures are inherently three-dimensional, plate-type two-dimensional (2D) shell elements are commonly employed in structural analysis because the in-plane dimensions are much larger than the thickness, making them effectively thin-walled structures. Since conventional structural integrity assessments are primarily based on applied stresses, shell element models are generally sufficient to capture structural behaviour under intact conditions. However, when potential flaws are to be considered, it becomes essential to determine the sectional stress distribution. In such cases, 2D shell elements have inherent limitations in accurately representing bending stresses, particularly in regions with geometric discontinuities where bending behaviour is dominant.

Several studies have applied Engineering Critical Assessment (ECA) to evaluate structural integrity, as summarised in Table 1. For example, Radu et al. (2020) conducted an ECA of weld joints in antenna tower structures according to the BS 7910 standard [11]. Rezaie et al. (2022) conducted an ECA of snake-laid pipelines using elastic–plastic finite element analysis [12], while He et al. (2023) applied ECA to monopile foundations of offshore wind turbines through the extended finite element method (XFEM) [13]. In these studies, structural analyses were conducted to calculate the stress intensity factor (SIF) using three-dimensional solid element models.

**Table 1.** Representative studies applying ECA to flaw acceptability analyses.

| Authors | Structure | Stress Intensity Factor Calculation | FE Analysis | Fracture Assessment |
|---|---|---|---|---|
| Radu et al. [11] | Antenna tower | Handbook (BS 7910) | Solid | O |
| He et al. [13] | Wind turbine monopile | XFEM | Solid | O |
| Seo et al. [14] | Type B (LNG) | Handbook (BS 7910) | Shell | O |
| Kim et al. [15] | Type C ($LH_2$, single) | Handbook (BS 7910) | Shell | - |
| Kim et al. [16] | Type C ($LCO_2$, single) | Handbook (BS 7910) | Shell | - |
| Present study | Type C ($LCO_2$, bilobe) | Handbook (BS 7910) | Shell/Solid | O |

The application of Engineering Critical Assessment (ECA) in ship structures commonly utilises two-dimensional shell element models. This is evidenced by studies such as the short-term and long-term crack propagation analyses of an IMO Type B cargo tank (Seo et al., 2023) [14] and the crack propagation analysis of an IMO Type C liquefied hydrogen cargo tank (Kim et al., 2024) [15]. However, research focusing on liquefied $CO_2$ ($LCO_2$) systems reveals a research gap: while structural analysis of $LCO_2$ tanks has been carried out based on the IGC Code (Kim et al., 2025) [16], ECA was not considered in that specific work. Recent ECA-based design studies for Type-C $LCO_2$ tanks have highlighted a critical issue: the use of high-strength steels introduces fracture sensitivity due to ductile-to-brittle transition and limited regulatory coverage within the current IGC

Code [17]. These findings demonstrated that fracture toughness requirements must be quantified to ensure sufficient safety margins when plate thickness and operating pressure increase. Furthermore, two key limitations exist in the existing research. First, although dynamic fracture behaviour in $CO_2$ transport systems has been investigated in pipeline environments [18], these studies do not address fracture evaluation or crack stability under the specific loading conditions of liquefied gas tanks on ships. Additionally, existing research has primarily focused on conventional cylindrical Type C tanks, with limited attention paid to the bilobe-type configuration. This configuration introduces geometric discontinuities at the web and intersection regions, which can lead to stress concentration and potential fracture sensitivity.

Despite the increasing application of Engineering Critical Assessment (ECA) for welded liquefied gas tanks, the effect of finite element modelling strategy—particularly the choice between shell and solid element formulations—on the stress linearisation process and subsequent FAD-based fracture evaluation has not yet been quantitatively clarified. Shell models are widely adopted in practice due to their computational efficiency; however, their simplified representation of through-thickness stress may introduce bias when assessing fracture reliability in geometrically complex regions, such as bilobe intersections and welded joints. Therefore, the objective of this study is to provide a systematic comparison of shell and solid finite element formulations within the BS 7910-based ECA framework for a bilobe IMO Type-C $LCO_2$ cargo tank. By quantifying the resulting differences in membrane and bending stresses, fracture ratio, load ratio, and safety factor, this work provides practical guidance on model selection and interpretation for assessing the structural integrity of emerging liquefied gas containment systems.

In the present study, a structural integrity assessment of a bilobe IMO Type C liquid $CO_2$ tank is performed using the flaw assessment methodology (BS 7910) proposed by the British Standards Institution. The remainder of this paper is structured as follows. Section 2 provides the background of the study of the Engineering Critical Assessment (ECA) framework. Section 3 describes the finite element modelling methodology, stress linearisation procedure, and fracture and fatigue assessment approach. Section 4 presents the results of the stress comparison and Engineering Critical Assessment for the shell and solid element models. Section 5 discusses the interpretation of the assessment outcomes. Finally, Section 6 summarises the key findings and concludes the study.

## 2. Research Background

### 2.1. Engineering Critical Assessment (ECA)

It is widely recognised that potential flaws, such as cracks, are unavoidable predecessors of structural failure. Crack propagation generally proceeds through three stages: initiation, stable growth under service conditions, and final fracture. Fatigue crack growth and fracture toughness are critical material properties for evaluating structural safety under extreme and cyclic loading conditions. Fracture toughness characterises the resistance of materials to crack growth and unstable propagation under monotonic loading and is typically evaluated through the J-integral or the crack tip opening displacement (CTOD). For fatigue loading, crack growth behaviour is generally described by the Paris Law, which relates the crack growth rate to the applied cyclic load. These fracture mechanics parameters provide a framework for understanding stable crack growth and the conditions leading to final fracture. Engineering Critical Assessment (ECA) embeds fracture mechanics principles with structural analysis to evaluate the acceptability of potential defects. Unlike conventional structural analysis, which considers only applied stress as the assessment criterion, fracture mechanics incorporates crack characteristics as additional variables alongside loading conditions. By integrating material resistance, crack dimensions, and applied loads, ECA enables a quantitative assessment of failure risk and supports strategies for designing and maintaining reliable structures [19]. This methodology has been widely applied in pressure vessels, ship structures, cryogenic tanks, and other structures where quantifying safety margins against flaws under fatigue and extreme loading is essential.

Engineering Critical Assessment (ECA) is commonly performed using standards based on the Failure Assessment Diagram (FAD) methodology. Several internationally recognised fitness-for-service and flaw assessment procedures are based on the FAD. One of the earliest frameworks is the R6 method, originally developed for the British electric power industry, which provides both simplified closed-form FAD options and a more advanced elastic-plastic approach using the J-integral. The BS 7910, developed by the British Standards Institution (BSI), and the FITness-for-service NETwork (FITNET), established by the EU, extended the method to welded structures and general industrial applications. In parallel, the API 579 standard offers a broader ECA framework that covers various flaw types and includes an extensive library of stress intensity factor solutions. Although recent revisions of R6 and BS 7910 introduced approaches to account for crack-tip constraint effects, their applicability to complex welded geometries remains limited, and conservative assumptions are typically adopted in practice [19].

### 2.2. Failure Assessment Diagram (FAD)

Within fracture mechanics-based flaw assessment methods, the Engineering Critical Assessment (ECA) employs the Failure Assessment Diagram (FAD). The FAD is the most widely used to analyse the elastic-plastic fracture behaviour of structures [19]. While structures may fail due to fracture by potential cracks, plastic collapse under overload can also occur in the absence of cracks. Steel materials exhibit ductile-to-brittle transition (DBTT) behaviour, being ductile at room temperature but brittle at low temperatures. Since liquefied gas tanks in ships are subjected to both room and cryogenic temperatures during loading and unloading, flaw assessments should be performed using elastic-plastic fracture mechanics. The FAD provides a framework for evaluating structural safety by representing the interaction of applied load and crack severity on two axes, as shown in Figure 1. The vertical axis defines the fracture ratio ($K_r$), and the horizontal axis defines the load ratio ($L_r$). The structural limit, as expressed by the failure assessment line (FAL), represents the relationship between the applied load and crack severity, which is influenced by material properties and geometry. Although numerical structural analyses (e.g., finite element method) can define location-specific FALs by considering elastic–plastic material behaviour, such analyses are computationally expensive. As an alternative, conservative empirical FALs, which are independent of geometry, are provided in standards such as BS 7910, and can be derived from the tensile test data. The empirical FAL formula is shown in Equation (1) [8].

$$\begin{aligned} f(L_r) &= \left(1+\tfrac{1}{2}L_r\right)^{-\frac{1}{2}}\left[0.3+0.7e^{-\mu L_r^6}\right], && \text{for } L_r \leq 1 \\ f(L_r) &= f(1)L_r^{\frac{N-1}{2N}}, && \text{for } 1 < L_r < L_{r,\max} \\ f(L_r) &= 0, && \text{for } L_r \geq L_{r,\max} \\ \text{where} & & & \\ \mu &= \min\left(0.001\tfrac{E}{\sigma_Y}, 0.6\right) & & \\ N &= 0.3\left(1-\tfrac{\sigma_Y}{\sigma_u}\right) & & \end{aligned} \quad (1)$$

The fracture ratio ($K_r$) corresponds to crack severity and is obtained by normalising the stress intensity factor ($K_I$) with respect to the fracture toughness ($K_{mat}$). $K_I$ quantifies the crack-tip singularity under remote loading and is typically evaluated for Mode I crack opening mode, which is induced by in-plane loading conditions. In the BS 7910 standard, $K_I$ is calculated using Equation (2), where load components ($P_m$, $P_b$, $Q_m$, $Q_b$) are decomposed and combined with stress magnification factors ($k_t$, $k_{tm}$, $k_{tb}$, $k_m$) [8].

$$\begin{aligned} K_I &= (Y\sigma)\sqrt{\pi a} \\ Y\sigma &= (Y\sigma)_p + (Y\sigma)_s \\ (Y\sigma)_p &= Mf_w\{k_{tm}M_{km}M_mP_m + k_{tb}M_{kb}M_b[P_b + (k_m - 1)P_m]\} \\ (Y\sigma)_s &= M_mQ_m + M_bQ_b \end{aligned} \tag{2}$$

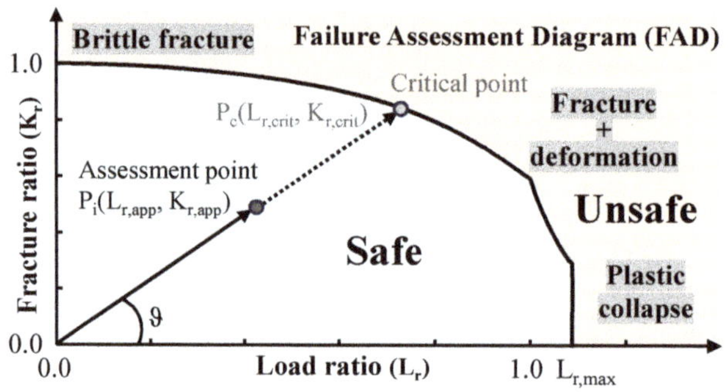


**Figure 1.** Configuration of the Failure Assessment Diagram (FAD).

The load ratio ($L_r$) corresponds to the severity of the applied load and is calculated by normalising the reference stress ($\sigma_{ref}$) with the yield stress ($\sigma_Y$). The reference stress is introduced particularly when the FAL is defined empirically, to account for geometry dependence effects. It is calculated using Equation (3), which reflects reductions in the effective section area due to the presence of cracks [8].

$$\begin{aligned} \sigma_{ref} &= \frac{P_b\left[P_b^2 + 9P_m^2(1-\alpha'')^2\right]^{0.5}}{3(1-\alpha'')^2} \\ &\text{where} \\ \alpha'' &= \frac{a/B}{1+B/c}, \qquad \text{for } W \geq 2(c+B) \\ \alpha'' &= (a/B)(2c/W), \qquad \text{for } W < 2(c+B) \end{aligned} \tag{3}$$

The mechanical behaviour of a flawed structure can be evaluated by the relative positions of the assessment point with respect to the failure assessment line (FAL), as illustrated in Figure 1. When the assessment point lies within the FAL, the structure is considered to be capable of sustaining the crack. In this case, the crack-related safety factor ($F_L$) can be determined from the distance ratio between the assessment point ($P_i$) and the critical point ($P_c$). Conversely, if the assessment point falls outside the FAL, structural failure is expected. The governing failure mechanism can then be distinguished by the relative magnitudes of the fracture ratio ($K_r$) and load ratio ($L_r$): fracture dominates when $K_r$ is critical, while plastic collapse occurs when $L_r$ is critical. In the intermediate region, failure may occur through fracture followed by plastic deformation. This behaviour can be quantitatively characterised by defining the degree of the assessment point ($\vartheta$). The safety factor and the degree are calculated using Equations (4) and (5), respectively.

$$F_L = \frac{P_c}{P_i} = \frac{K_{r,crit}}{K_{r,app}} = \frac{L_{r,crit}}{L_{r,app}} \tag{4}$$

$$\theta = \arctan\left(\frac{K_{r,app}}{L_{r,app}}\right) = \arctan\left(\frac{K_{r,crit}}{L_{r,crit}}\right) \tag{5}$$

Fatigue damage in structures can be described as the growth of an initial crack to a critical size, ultimately leading to fracture. The crack growth rate ($\Delta a$) is expressed as a function of the stress intensity factor range ($\Delta K_I$). As illustrated in Figure 2, fatigue crack growth behaviour is generally divided into three stages. When $\Delta K_I$ is below the threshold stress intensity factor ($K_{th}$), which defines the minimum condition for crack propagation, no crack growth occurs. When $\Delta K_I$ exceeds the critical value ($K_{Ic}$ or $J_{Ic}$), fracture occurs prior to significant crack growth. In the intermediate region, the crack growth rate follows an exponential relationship with $\Delta K_I$, commonly expressed by the Paris Law (Equation (6)), which appears as a straight line on a logarithmic plot [19].

$$\frac{da}{dN} = C\Delta K^m \tag{6}$$

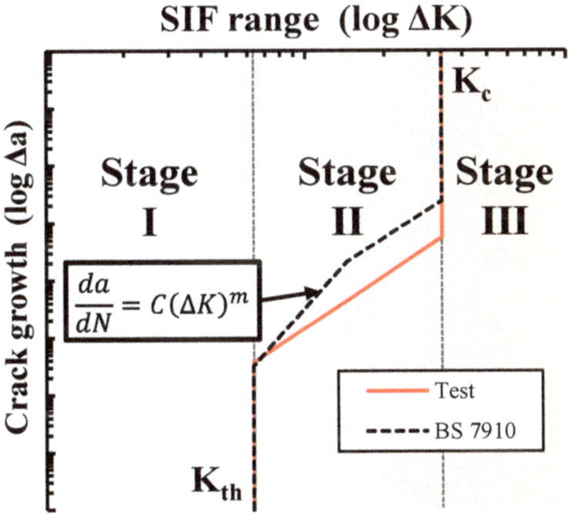


**Figure 2.** Fatigue crack growth behaviour.

The fatigue crack growth coefficients ($C$, $m$) vary with material and can be obtained either from standardised testing or from recommended values in ECA procedures. In this study, both experimental data from ASTM E647 tests and recommended values from the BS 7910 ECA standard were applied and compared [20,21].

### 2.3. Finite Element Configurations

The sectional stress distribution required for the ECA procedure in finite element analysis (FEA) depends on the configuration of integration points. Therefore, it is essential to understand the characteristics of shell and solid elements in order to analyse stress distribution at assessment locations. These characteristics are primarily determined by the underlying theoretical formulations, which define stress integration through the thickness.

The proportion of stress components also varies with the type of applied loading, ultimately influencing the results of ECA.

Thin-walled structures are usually modelled with two-dimensional elements in structural analysis. When representing three-dimensional structures using two-dimensional surfaces, curved geometries must be included. Accordingly, shell elements are employed to simulate three-dimensional behaviour based on plate theories. Shell elements can be divided into thin and thick formulations. Thin shell elements are based on the Kirchhoff–Love plate theory, which is analogous to Euler–Bernoulli beam theory and neglects shear deformation through the thickness. In contrast, thick shell elements adopt the Reissner–Mindlin plate theory, which is analogous to Timoshenko beam theory and accounts for transverse shear deformation. In this formulation, cross-sections are assumed to rotate due to shear effects while maintaining planarity. In the finite element software ABAQUS 2019, general-purpose shell elements are based on the Reissner–Mindlin theory. Because plate theories explicitly include bending moments as external effects, both nodal forces and nodal moments are considered in shell elements. Furthermore, since thickness is treated as a stiffness parameter rather than as part of the geometry, shell elements offer advantages such as convenient design modifications and simplified modelling of thickness transition regions and cruciform joints in welded structures.

Solid elements, on the other hand, are three-dimensional volumetric elements based on continuum mechanics. In this formulation, only external forces are considered without explicit representation of nodal moments. Since thickness is represented as a true geometric dimension, solid elements enable accurate modelling of thickness variations and joint connections. However, design modifications require dimensional changes, which increase modelling costs and complexity compared with shell elements.

## 3. Research Methodology

The workflow for comparing Engineering Critical Assessment (ECA) results based on shell and solid element models is summarised in Figure 3.

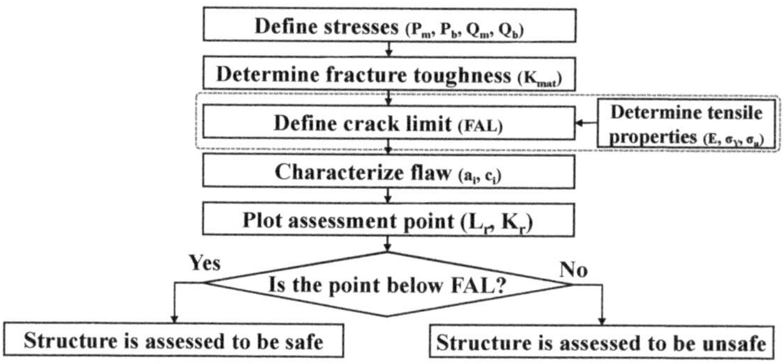


**Figure 3.** Flaw severity assessment procedure by Engineering Critical Assessment.

### *3.1. Applied Structure*

#### 3.1.1. Tank Specification

For pressurised ship cargo tanks with a design pressure exceeding 2 atm, the IMO Type C configuration is applied. While Type C tanks are generally designed with a cylindrical shape, in this study, the liquefied $CO_2$ ($LCO_2$) tank is designed in a bilobe shape to enhance storage efficiency. The tank features a dome on the topside and a sump on the bottom side, enabling efficient and safe bunkering and maintenance. However, these attached structures introduce geometric discontinuities that cause local stress concentration. Additionally, the bilobe shape necessitates a Y-joint to connect two cylinders, which introduces an additional structural vulnerability.

The target $LCO_2$ tanks have a total volume of approximately 9800 m$^3$, with an internal diameter of about 15.6 m and a length of 36 m The tank is supported longitudinally by two saddle structures. To accommodate thermal expansion and contraction, the forward support is designed as a sliding support, allowing axial movement. Internally, a bulkhead and webs are installed to provide self-supporting capacity and to reduce sloshing-induced loads, with stiffeners arranged to reinforce these structures. For the ECA, assessment locations are selected at the weld joints of the web, dome, and sump, as shown in Figure 4. In addition, regions with thickness transitions at welds are also considered assessment locations to account for the influence of geometric discontinuities.

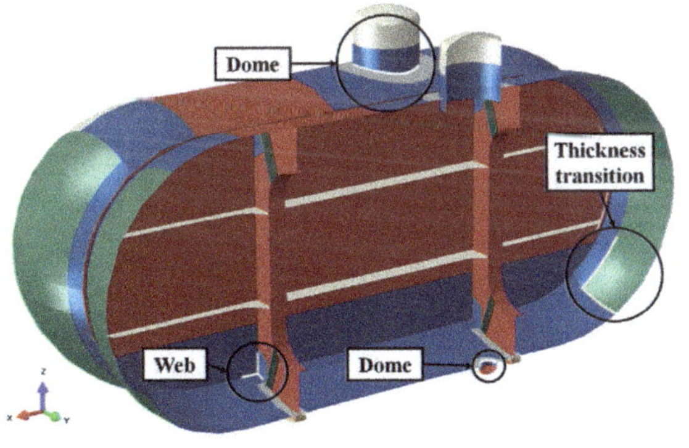


**Figure 4.** Target structure and assessment locations.

### 3.1.2. Material Properties

The tank and its supports were fabricated from LTFH36 steel, with wood inserted between the tank and the saddle supports. The material properties of the steel and wood used in the structural analysis are summarised in Table 2.

**Table 2.** Material properties of LFTH36 steel and wood.

| Material Properties | LTFH36 | Supporting Wood |
|---|---|---|
| Elastic Modulus (MPa) | 206,000 | 30,000 |
| Poisson Ratio | 0.3 | 0.4 |
| Yield Stress (MPa) | 355 | - |
| Tensile Strength (MPa) | 490 | - |
| Density (kg/m$^3$) | 7850 | 1300 |
| Thermal Expansion Coefficient ($\times 10^{-6}$ °C) | 12 | 8 |

For conducting the Engineering Critical Assessment (ECA), fracture toughness and fatigue crack growth rate (FCGR) are essential material properties. Fracture toughness can be determined through standardised testing methods such as ASTM E1820, BS 7448, or ISO 12135 [22–24]. In this study, the fracture toughness of LTFH36 steel was determined in accordance with ASTM E1820, obtaining the J-integral. The FCGR of LTFH36 steel was determined in accordance with the ASTM E647 standard [25]. The ECA results, based on experimentally obtained data, were compared with those using the recommended FCGR data provided in the BS 7910 standard. The measured fracture toughness of LTFH36 was

333 MPa$\sqrt{m}$, and the FCGR data are presented in Table 3. The test equipment and specimen is shown in Figure 5.

**Table 3.** Fatigue crack growth data.

| Curve | Threshold ($N/mm^{3/2}$) | A (Mean + 2SD) | | m (Mean + 2SD) | | Transition ($N/mm^{3/2}$) |
|---|---|---|---|---|---|---|
| | | Stage A | Stage B | Stage A | Stage B | |
| Recommended (Weld) | 63 | $2.10 \times 10^{-17}$ | $1.19 \times 10^{-12}$ | 5.10 | 2.88 | 144 |
| Experiment | 63 | $1.49 \times 10^{-13}$ | | 3.0 | | - |

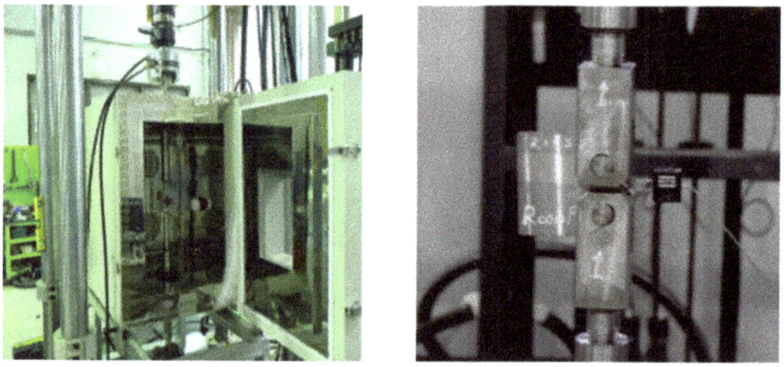

(a) (b)

**Figure 5.** Test equipment and specimen. (**a**) Test equipment; (**b**) Test specimen for FCGR test.

Accurate fracture toughness values are essential for the reliability of ECA, particularly in welded structures where local constraint and geometric discontinuities are pronounced. Recent studies have shown that the measure of toughness can vary significantly depending on specimen thickness and crack orientation, with reductions of up to approximately 70% reported in weld and heat-affected zones (HAZs) [26]. Such findings highlight material and weld-specific fracture toughness data when assessing $LCO_2$ cargo tanks, where the bilobe web connections and associated welds act as stress concentration locations.

### 3.1.3. Loading Conditions

The loading components considered in the structural analysis of the gas tank include acceleration, pressure, and thermally induced loads. Acceleration can be categorised into static and dynamic components. Static acceleration corresponds to gravity (self-weight), whereas dynamic acceleration arises from ship motion. In this study, the three-dimensional components of dynamic acceleration were adopted from the recommended values in the IGC Code, which are defined at a probability level of $10^{-8}$ for the North Atlantic Sea. These values are calculated based on the principal dimensions of the cargo ship and the centre of gravity of the tank. The recommended formulae for calculating acceleration are provided in Equation (7), and the principal dimensions of the ship and tank are summarised in Table 4 [6].

$$
\begin{aligned}
a_z &= \pm a_0\sqrt{1+\left(5.3-\frac{45}{L_0}\right)^2\left(\frac{x}{L_0}+0.05\right)^2\left(\frac{0.6}{C_B}\right)^{1.5}+\left(\frac{0.6yK^{1.5}}{B_s}\right)^2} \\
a_y &= \pm a_0\sqrt{0.6+2.5\left(\frac{x}{L_0}+0.05\right)^2+K\left(1+0.6K\frac{z}{B}\right)^2} \\
a_x &= \pm a_0\sqrt{0.06+A^2-0.25A} \\
a_0 &= 0.2\frac{V}{\sqrt{L_0}}+\frac{34-600/L_0}{L_0} \\
\text{where} & \\
a_0 &= 0.2\frac{V}{\sqrt{L}}+\frac{34-\frac{600}{L}}{l} \\
A &= \left(0.7-\frac{L}{1200}+5\frac{z}{L}\right)\left(\frac{0.6}{C_B}\right)
\end{aligned}
\tag{7}
$$

**Table 4.** Principal dimensions of $LCO_2$ carrier and gas tanks.

| $L_0$ (m) | $B_s$ (m) | $C_B$ | $V$ (knot) | $x$ (m) | $y$ (m) | $z$ (m) |
|---|---|---|---|---|---|---|
| 215.0 | 31.6 | 0.78 | 14 | 53.5 | 0.0 | 2.8 |

The load cases for the fracture and fatigue assessments consist of three dynamic acceleration load cases during voyage (LC1–LC3) and one static load case during loading and unloading (LC4). In addition, an accident load case (LC5) was considered, corresponding to a collision scenario with inertial accelerations of 0.5 g in the forward direction and 0.25 g in the aft direction, as recommended by the IGC Code [6]. These load cases are presented in Table 5. In these load cases, the internal pressure ($P_0$) and hydrostatic pressure were applied to the inner surface of the tank shell, while gravity and dynamic acceleration load components were introduced as body forces acting on both the tank shell and the supporting woods. In addition, thermal deformation due to the temperature difference was applied to the tank, excluding the supporting woods and saddles.

**Table 5.** Load cases considered for structural analysis.

| Load Case | Internal Pressure | Static Loading | Acceleration | Temperature |
|---|---|---|---|---|
| Longitudinal ($x$, LC1) | $P_0$ | hydrostatic pressure, gravity | $\pm a_x$ | $T_{cargo}$ |
| Transverse ($y$, LC2) | $P_0$ | hydrostatic pressure, gravity | $\pm a_y$ | $T_{cargo}$ |
| Vertical ($z$, LC3) | $P_0$ | hydrostatic pressure, gravity | $\pm a_z$ | $T_{cargo}$ |
| Static (LC4) | - | gravity | - | $T_{room}$ |
| Collision (LC5) | $P_0$ | hydrostatic pressure, gravity | +0.5 g, −0.25 g | $T_{cargo}$ |

The characteristics of cyclic loading during the design life should be defined for fatigue assessment. For liquefied gas carriers, fatigue loading can generally be categorised into high-cycle fatigue (HCF) and low-cycle fatigue (LCF). HCF corresponds to the loading induced by ship motions under environmental conditions during voyages. In this study, the design life of the $LCO_2$ carrier is assumed to be 20 years, corresponding to approximately $10^8$ cycles when the zero-crossing frequency is taken as 1/7 Hz, as provided in Equation (8).

$$
n_d = \nu_0 T_d = \frac{1}{7}\text{Hz} \times (20 \times 365.25 \times 24 \times 3600)\ \text{seconds} = 10^{7.955} \approx 10^8 \tag{8}
$$

Since fatigue loading on the tanks is negligible when the cargo is empty, a voyage factor, defined as the ratio of the loaded state, is assumed to be 0.5. Furthermore, because dynamic loads from the marine environment are absent when the ship is anchored, a harbour factor, defined as the ratio of the harbour state, is assumed to be 0.15. The load distribution during HCF is represented by a Weibull distribution with a shape parameter of 1. The total number of HCF loadings, without applying voyage and harbour factors, is $10^8$. The corresponding probability density function and rank-size distribution are shown in Figure 6.

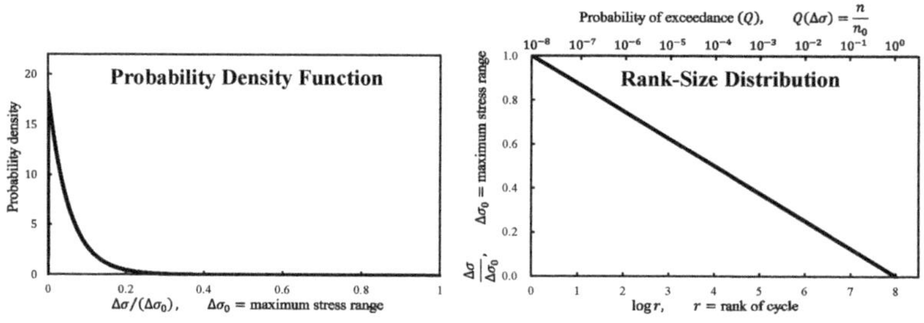


**Figure 6.** Probability density function and rank-size plot for the Weibull distribution.

LCF, on the other hand, is associated with cargo loading and unloading operations in the harbour. The LCF load distribution is assumed to be constant, and the number of cycles depends on the operational schedule, which governs the average voyage period. In this study, the number of LCF cycles is assumed to be 1000, which is the minimum recommended value for a 20-year design life according to the IGC Code.

### *3.2. Finite Element Modelling*

#### 3.2.1. Shell and Solid Element

Based on the selected principal dimensions of the structures, the liquefied gas tank was modelled using both shell and solid elements for structural analysis. In this study, a nonlinear finite element analysis was performed using the structural analysis software Abaqus 2019, which is capable of accounting for large deformations, including contact between the tank and supports, as well as thermally induced deformation. For the shell element model, the S4R element was employed. This element is based on thick plate theory and uses reduced integration to prevent shear locking. Considering the maximum plate thickness (50 mm), the in-plane mesh size (width, length direction) is set to 150 mm, and five integration points were assigned through the thickness to compute forces and moments.

For the solid element model, the C3D8R element, a reduced-integration continuum element, is employed. Since plate thickness should be explicitly modelled in solid elements, weld beads are applied at thickness change regions and intersections, as shown in Figure 7. The mesh size in the width and length directions is also set to 150 mm, while the thickness direction is divided into four elements across the section. This configuration ensured that the number of integration points through the thickness in the shell model corresponds to the number of nodes in the solid model. The mesh layouts of both the shell and solid element models are illustrated in Figure 8.

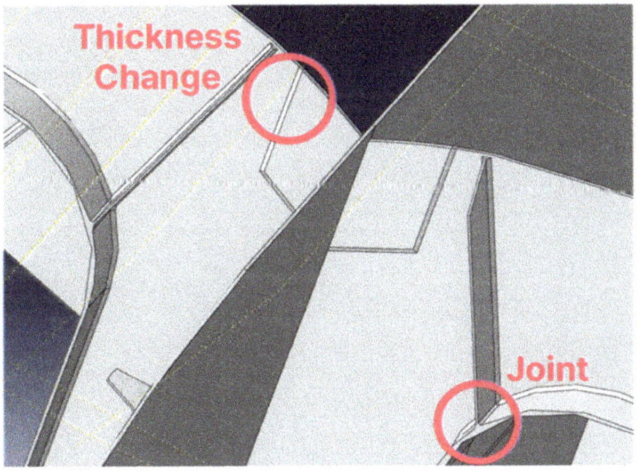


**Figure 7.** Geometries of weld beads in solid model.

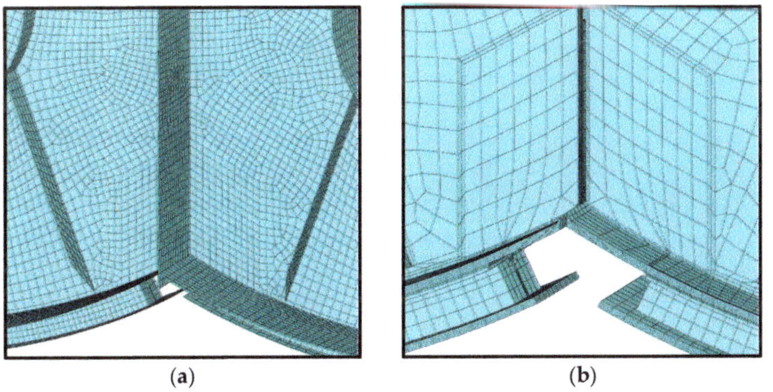

(**a**) (**b**)

**Figure 8.** Mesh configurations of the shell and solid model. (**a**) Shell element model; (**b**) Solid element model.

3.2.2. Stress Linearisation

In accordance with BS 7910, where the load ratio and fracture ratio are defined in terms of the membrane ($P_m$, $Q_m$) and bending ($P_b$, $Q_b$) stress components (Equations (2) and (3)), the through-thickness stress distribution was linearised into membrane and bending stresses. The membrane stress corresponds to the mean stress across the thickness, and the bending stress represents the linearly varying component. For shell elements, the stresses at the bottom (SNEG, $\sigma_{bottom}$) and the top (SPOS, $\sigma_{top}$) obtained from the structural analysis are utilised, and the membrane stress ($P_m$) and the bending stress ($P_b$) are calculated using Equation (9).

$$\begin{aligned} P_m &= \frac{\sigma_{bottom} + \sigma_{top}}{2} \\ P_b &= \frac{|\sigma_{bottom} - \sigma_{top}|}{2} \end{aligned} \quad (9)$$

For solid elements, the target section for linearisation shall first be defined. If the section coincides with element edges, stresses can be calculated at (number of elements + 1) locations. Otherwise, when the section intersects element interiors, stresses at the corresponding positions must also be evaluated. In this study, stress linearisation is performed using the Abaqus structural analysis software. After defining the assessment path, it is divided into 40 locations where stresses were extracted. The stress distribution is then linearised by fitting a first-order polynomial. An example of stress linearisation for the solid element model is shown in Figure 9.

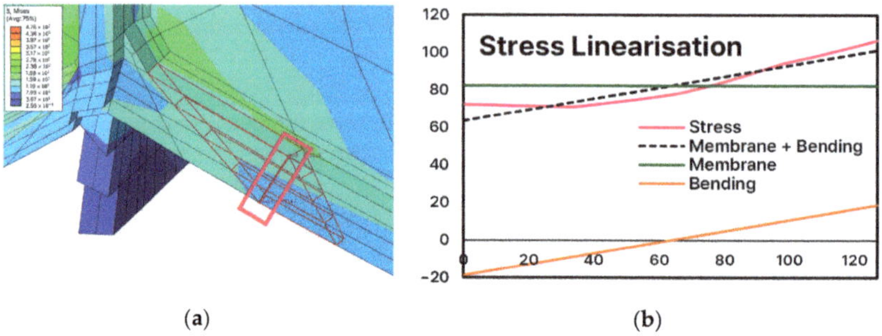


**Figure 9.** Stress linearisation for the solid element model. (**a**) Define assessment path; (**b**) Stress linearisation result.

### 3.3. *Engineering Critical Assessment*

#### 3.3.1. Flaw Characterisation

The flaw geometry adopted in this study is illustrated in Figure 10. Although the $LCO_2$ tank is designed with a curved cylindrical geometry, it is assumed to be a flat plate because the thickness-to-diameter ratio ($t/D$) is less than 0.1. A surface flaw is considered, as it produces a higher stress intensity factor than an embedded flaw. The initial crack size can be determined by the detection capability of non-destructive testing (NDT). In this study, the initial dimensions were assumed to be $a$ = 6 mm and $2c$ = 25 mm, corresponding to the minimum detectable size by ultrasonic testing (UT) as recommended in the BS 7910 standard.

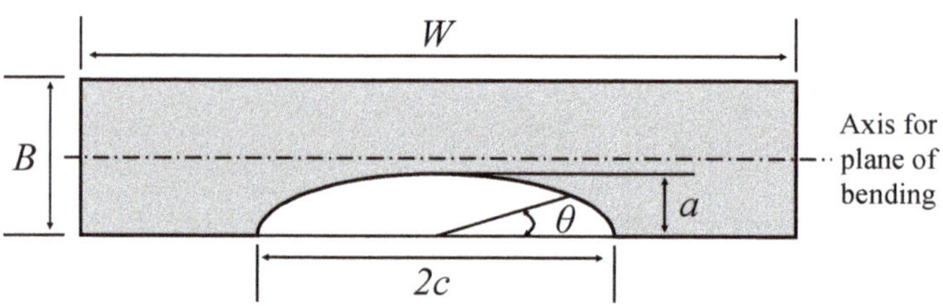


**Figure 10.** Geometry of surface flaw. The figure illustrates the plate dimensions $W$ and $B$, the flaw with depth $a$ and half length $c$, and the crack-front angle $\theta$. The dashed line represents the neutral axis for bending.

For the shell element model, stress magnification factors at the weld toe and root are taken into account. The geometries of the butt weld and cruciform joint are shown in Figure 11, with the overall attachment length between weld toes defined as $L$ = 80 mm.

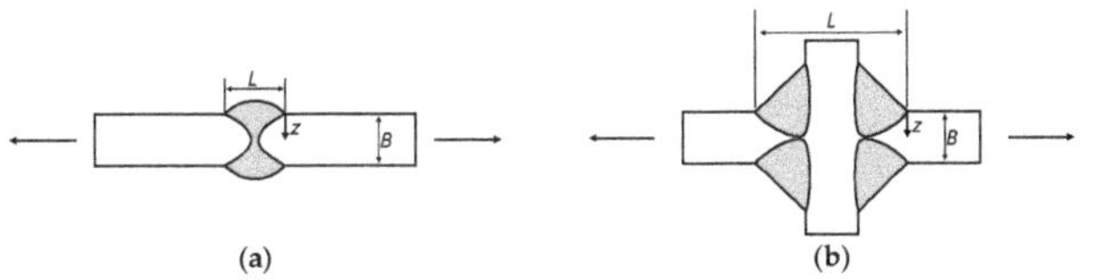


**Figure 11.** Geometries of weld joints. (**a**) Butt weld; (**b**) Cruciform joint. In both cases, $z$ is the height measured from the weld, and $L$ is the overall length from weld toe to weld toe.

Fabrication-induced imperfections caused by welding misalignment were also considered. Misalignment can be categorised as angular or axial, both of which introduce additional bending moments under axial loading. In this study, only butt joint misalignments are considered, as they are unavoidable in thickness transition regions. The types of misalignments at butt joints are shown in Figure 12, where the angular misalignment is assumed to be $d$ = 5 mm and the axial misalignment $e$ = 3 mm.

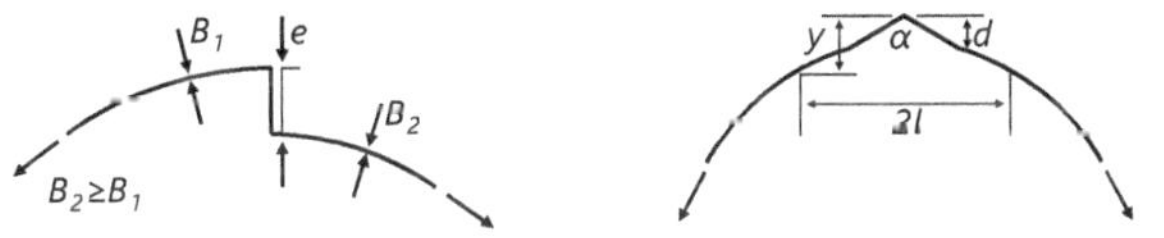


**Figure 12.** Misalignment in butt joints. (**a**) Axial misalignment, where $B_1$, $B_2$, and e represent the plate thicknesses and the axial misalignment, respectively; (**b**) Angular misalignment, where d, y, α and l represent the deviation from true circle, the peaking height, and the angular change due to misalignment.

3.3.2. Fracture Assessment Methodology Using ECA

The fracture assessment by the Engineering Critical Assessment (ECA) is performed using an initial crack size of $a$ = 6 mm and $2c$ = 25 mm. The applied loading conditions are defined at a probability level of $10^{-8}$ for the North Atlantic Sea, corresponding to load cases LC1–LC3 (dynamic) and LC4 (static). The assessment points are then plotted on the Failure Assessment Diagram (FAD), and the results of the shell and solid element models are evaluated and compared in terms of the safety factor ($F_L$) and the ratio of $K_r$ to $L_r$ ($\vartheta$).

3.3.3. Fatigue Assessment Methodology Using ECA

The fatigue assessment by the ECA is performed in two steps. First, the final crack dimensions ($a$, $2c$) after fatigue loading are calculated. In this step, only the fatigue crack growth rate of the material is considered; fracture due to the stress intensity factor (SIF) exceeding the material resistance ($K_{mat}$) is not taken into account. If the final crack depth ($a_f$) exceeds half of the plate thickness ($0.5t$), the flaw is assessed as a through-thickness crack, which is regarded as unacceptable. The final crack depths obtained from the shell and solid element models are compared. To evaluate the influence of crack growth law data, both the recommended values from the BS 7910 standard and the experimental data are applied and compared.

Second, a fracture assessment is performed based on the final crack size. Since $K_r$ primarily increases with crack length, while $L_r$ slightly increases due to the reduction in sectional area, the assessment point shifts toward the failure assessment line (FAL). Thus, a structure deemed safe at the initial crack size may become unsafe at the final stage. The results of the shell and solid element models were evaluated and compared in terms of the safety factor ($F_L$) and the ratio of $K_r$ to $L_r$ ($\vartheta$).

# 4. Results

## 4.1. Stress Distribution

The structural analysis information of the shell element model with five integration points through the thickness and the solid element model with four elements through the thickness, both of which were used for the final structural integrity assessment, is summarised in Table 6. It was observed that although the solid element model involved approximately 2.78 times more elements, the computation time required was about 4.78 times longer.

**Table 6.** Summary of Structural Analysis of Shell and Solid Element Models.

| Analysis Information | Shell Element | Solid Element |
|---|---|---|
| Number of Elements | 372,999 | 1,037,272 |
| Number DOFs | 2,182,674 | 3,314,808 |
| CPU Time (s) | 34,681 | 165,847 |

To examine the variation of stress components with thickness, the membrane and bending stresses of the shell and solid element models were calculated at three thickness locations ($t$ = 30 mm, 40 mm, and 50 mm). The corresponding von Mises stress distributions are presented in Figures 13–15. The results are summarised in Tables 7–9. However, the membrane and bending stresses were obtained by linearising the principal stresses through the thickness, rather than using the von Mises stress directly. The membrane stress component tended to be up to approximately 1.38 times higher in the shell element model as the thickness decreased. Although the bending stress in the shell element model was also higher, there was no clear trend observed for the bending stress component.

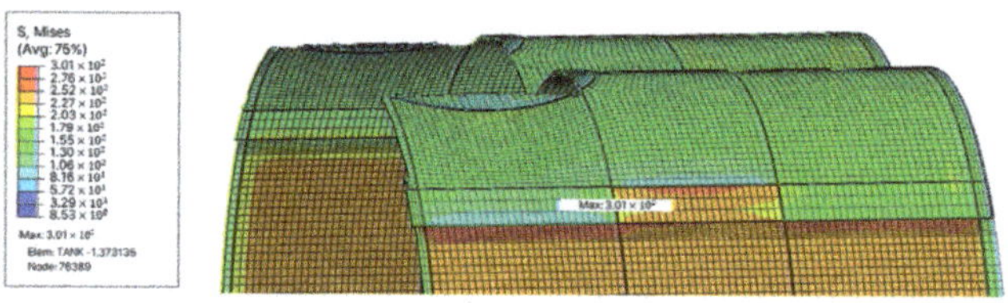


**Figure 13.** Von Mises stress distribution of tank side shell ($t$ = 30 mm).

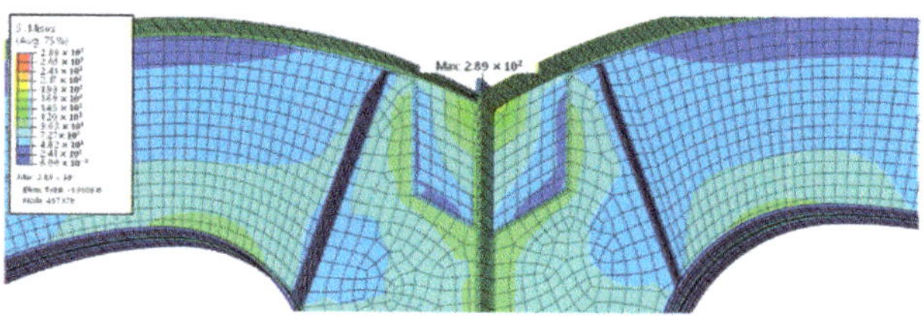


**Figure 14.** Von Mises stress distribution of tank topside shell ($t$ = 40 mm).

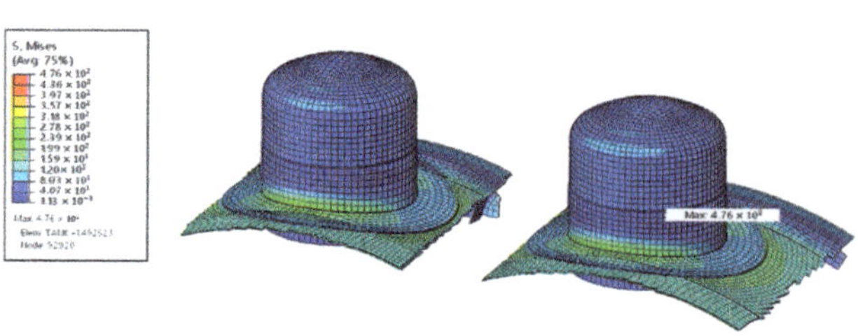


**Figure 15.** Von Mises stress distribution of tank dome ($t$ = 50 mm).

**Table 7.** Stress components for tank side shell ($t$ = 30 mm).

| Shell Element Model | | Solid Element Model | |
|---|---|---|---|
| **Membrane (MPa)** | **Bending (MPa)** | **Membrane (MPa)** | **Bending (MPa)** |
| 150.5 | 163.5 | 109.0 | 95.1 |

**Table 8.** Stress components for tank topside shell ($t$ = 40 mm).

| Shell Element Model | | Solid Element Model | |
|---|---|---|---|
| **Membrane (MPa)** | **Bending (MPa)** | **Membrane (MPa)** | **Bending (MPa)** |
| 266.0 | 96.7 | 238.8 | 93.7 |

**Table 9.** Stress components for tank dome ($t$ = 50 mm).

| Shell Element Model | | Solid Element Model | |
|---|---|---|---|
| **Membrane (MPa)** | **Bending (MPa)** | **Membrane (MPa)** | **Bending (MPa)** |
| 328.7 | 172.3 | 314.5 | 103.3 |

### 4.2. Fracture Assessment Results

The results of the fracture assessment for three dynamic load cases (LC1–LC3) and one static load case (LC4) are presented in Figure 16. Among the assessed locations, the weld joint at the shell is identified as the most vulnerable, followed by the web.

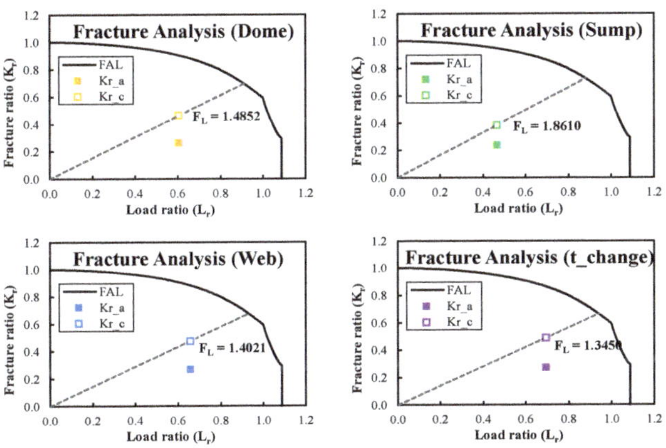


**Figure 16.** The fracture assessment results from the ECA. The dashed lines indicate the extension of the assessment point used for calculating safety factors.

### 4.3. Fatigue Assessment Results

The fatigue assessment results based on the ECA, which utilised the BS 7910 standard, are presented in Figure 17. At all assessment locations, the final crack depth did not exceed 0.5$t$, which is assumed to represent through-thickness penetration. However, significant variations in the final crack depth were observed depending on the crack growth rate properties applied in the analysis. In the sump, a minimum difference of 5.71 times was observed, whereas in the web, the difference reached 6.62 times.

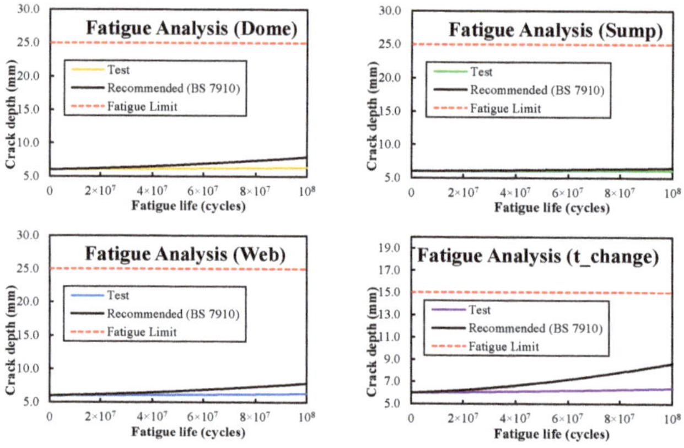


**Figure 17.** The fatigue assessment results by the ECA.

The fatigue assessment considering fracture, based on the ECA, is presented in Figure 18. The utilisation of fatigue crack growth rates derived from experiments indicated that the crack in the final state would not lead to failure. In the thickness transition region, when the crack growth rate recommended by the BS 7910 standard was applied, the enlarged crack did not result in fracture; however, the safety factor decreased from 1.3075 to 1.1958.

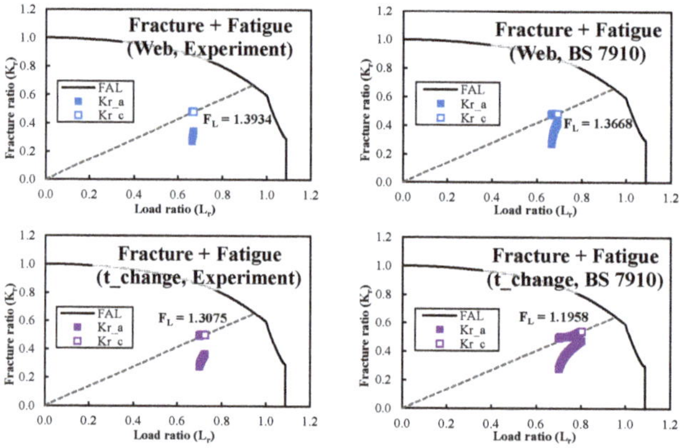


**Figure 18.** The fatigue assessment results combining the fracture by the ECA. The dashed lines indicate the extension of the assessment point used for calculating safety factors.

The safety factors calculated from the results were compared by plotting the values of the shell element model on the *x*-axis and those of the solid element model on the *y*-axis, as shown in Figure 19. The comparison showed that the safety factors of the solid element model were, on average, 1.12 times higher, with a maximum difference of 2.14 times. The coefficient of variation (Cov) was 0.088, and the correlation coefficient (*r*) was 0.482, indicating a moderate correlation.

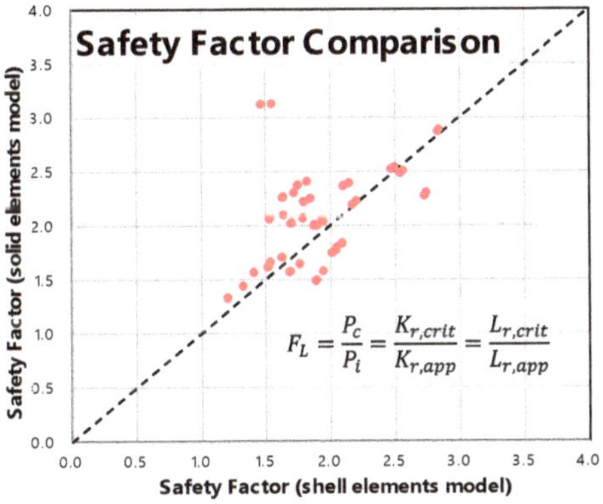


**Figure 19.** Safety factor comparison of shell and solid element models.

## 5. Discussion

The comparison between shell and solid element models reveals a balance between computational efficiency and stress prediction accuracy. The shell model effectively captures global structural behaviour with low cost but tends to overestimate membrane stresses due to simplified through-thickness representation. In contrast, the solid model provides more realistic stress distributions, particularly near welds and geometric transitions, albeit with significantly higher computational demand. These differences lead to variations in the Failure Assessment Diagram, where the solid model yields lower $K_r$ and $L_r$ values, indicating that accurate local stress evaluation is crucial for ECA in complex regions. Fatigue analysis revealed that, while overall crack growth trends were consistent across modelling approaches, the final crack depth was highly sensitive to the choice of fatigue crack growth rate (FCGR) dataset, varying by up to a factor of eight. This highlights that material properties, especially fracture toughness and FCGR, exert a greater influence on life prediction than the selection between shell and solid modelling.

The higher fracture risk identified in the web and Y-joint regions is associated with the bilobe tank, where curvature transition and weld-induced discontinuities amplify stress gradients. The solid model captured these localised effects more clearly, implying that a hybrid modelling strategy would provide an effective and practical approach for ECA of bilobe type C tanks. The FCGR datasets used in this study inherently represent different confidence levels within the statistical distribution of crack growth behaviour. While the shell and solid models yielded consistent qualitative trends, the sensitivity of the assessment outcome to FCGR selection underscores the role of material-data uncertainty in operational reliability. Therefore, extending the current deterministic ECA framework toward a probabilistic assessment—incorporating the statistical scatter of FCGR and fracture toughness—would provide a more robust basis for evaluating structural integrity and service life. This is identified as an important direction for future work. Fatigue crack growth predictions are highly dependent on material-specific parameters, supporting the need to consider FCGR dataset variability [27]. While this study focuses on the comparative ECA methodology for liquefied gas tanks, the results are relevant to classification society certification practices, where ECA may be used as an alternative acceptance route to replace or relax PWHT requirements for plate thicknesses up to 50 mm.

In this study, the initial crack size and the fatigue crack growth rate were treated as fixed values, and uncertainties arising from geometric imperfections and load spectra were not considered. Future work may incorporate probabilistic Engineering Critical

Assessment (ECA) to account for these uncertainties. Statistical modelling frameworks, such as the symmetric unimodal distribution–based approaches suggested in [28] or the realised PDF-based approach [29], offer a promising basis for extending the deterministic ECA methodology toward reliability-informed structural integrity assessment and digital–twin–based monitoring of $LCO_2$ tanks. Moreover, low-temperature fracture toughness exhibits substantial statistical scatter, as demonstrated in recent compilations of cryogenic fracture data [30].

This analysis was performed for a fixed–sliding support configuration as defined in the tank's original design. Since the sectional stress distribution—which is critical for stress linearisation in ECA—can vary depending on support and boundary conditions, future work may include a parametric evaluation of support schemes and boundary configurations to examine their influence on stress gradients and ECA results. In addition, the effect of tank supports subjected to high-cycle fatigue loading should also be considered in future assessments. The current assessment framework focuses on static and fatigue loading conditions. Incorporating vibration-induced stress effects represents a valuable direction for further development, particularly for evaluating structural integrity under dynamic operational environments. The reliability of the finite element models depends in part on the discretisation strategy; however, a comprehensive quantitative mesh sensitivity study was not conducted, as the primary objective of this work was to examine the comparative differences between shell and solid element formulations. To ensure the validity of this comparison, the same mesh sizing principles and stress linearisation procedures were applied consistently in both models, so that the relative trends observed remain robust.

For curved shell structures, particularly those containing geometric discontinuities such as bilobe web intersections and welded regions, local mesh refinement is necessary to accurately capture stress gradients and nonlinear deformations. Prior numerical studies recommend increasing element density in these curvature-transition regions to avoid underestimating local stress concentrations [31]. Nevertheless, the structural integrity assessment in this study is governed by the fracture ratio and load ratio, which are derived from the membrane and bending stress components. These linearised stress components are generally less sensitive to localised mesh variations than peak surface stresses. Thus, while local stress magnitudes may vary with refinement, the membrane–bending decomposition used for the Failure Assessment Diagram (FAD) remained stable between the two model types, reinforcing the reliability of the comparative conclusions drawn in this study.

The computational time difference between the shell and solid formulations is explicitly noted in the Results section. However, a definitive performance comparison based on the accuracy per unit of computational cost metric was not included. We recognise the importance of this metric for industrial adoption. Establishing absolute accuracy requires extensive experimental validation or a fully verified analytical solution, which was beyond the scope of this theoretical comparative study. Therefore, we propose to address this quantitative assessment in detail within a future work programme dedicated to experimental validation. The comparison of our simulation results focuses primarily on the internal consistency between the shell and solid finite element formulations. We recognise the absence of direct comparison with experimental verification or external scholarly findings. However, our Engineering Critical Assessment (ECA) framework and stress linearisation methodology rigorously comply with internationally recognised standards. This adherence ensures the methodological reliability of our approach. Obtaining external data that precisely matches our complex, specific geometry and loading conditions was not feasible; thus, validation relies on method compliance and demonstrating the systematic relative differences between the two modelling techniques.

## 6. Conclusions

In this study, an Engineering Critical Assessment was conducted for the bilobe IMO type C liquid $CO_2$ tank, and the results obtained from the shell and solid element models were systematically compared. The findings of the study can be summarised as follows.

- Accuracy and computational efficiency: The shell model required significantly fewer elements and reduced computation time, whereas the solid model increased computational cost by a factor of 4.78 to achieve improved local stress resolution.
- Stress distribution and $K_r$, $L_r$ distribution: Both models exhibited similar global stress behaviour, but the solid model captured sharper local stress gradients, influencing the membrane/bending stress components and subsequently the fracture ratio ($K_r$), load ratio ($L_r$), and safety factor ($F_L$).
- Fatigue and fracture assessment: While through-thickness crack growth was not predicted, the final crack depth showed sensitivity to FCGR datasets, indicating the importance of accurate crack growth and fracture toughness data for $LCO_2$ tank steels (ex., LTFH36).
- Characteristics by location: The web and welded intersections of the bilobe shell were consistently identified as critical locations due to geometric and weld-induced stress concentrations.

This study did not consider the effect of tank supports, which are subjected to high-cycle fatigue loading. Nevertheless, the comparative ECA procedure developed here provides valuable insights into the influence of finite element modelling approaches and is expected to serve as a reference for the structural analysis of liquefied gas tanks, particularly in welded regions. This study does not propose a new fracture assessment model; rather, its contribution lies in quantifying the model-form uncertainty associated with the shell and solid element formulations in ECA of bilobe $LCO_2$ tanks. The results show that while shell models are efficient for global integrity assessment, solid models provide improved stress resolution in critical welded and geometrically discontinuous regions. These findings provide a scientific basis for selecting FE modelling complexity according to the assessment purpose, balancing computational cost and the required local accuracy.

Future work will focus on incorporating mesh sensitivity quantification and probabilistic ECA, to account for the statistical scatter of fracture toughness and fatigue crack growth rate data, thereby enabling reliability-based service-life evaluation. Lastly, it is considered necessary to perform more detailed three-dimensional finite element analyses [32], in which the boundary conditions and internal loads are modelled as realistically as possible, to enable a more in-depth investigation.

**Author Contributions:** Conceptualisation, D.I.K., Y.Y.L. and D.K.K.; methodology, D.I.K.; validation, D.I.K., Y.Y.L., N.-K.C. and J.-H.H.; formal analysis, D.I.K.; investigation, N.-K.C.; resources, D.K.K.; data curation, N.-K.C.; writing—original draft preparation, D.I.K., Y.Y.L. and D.K.K.; writing—review and editing, N.-K.C., J.-H.H. and D.K.K.; visualisation, D.I.K.; supervision, D.K.K.; project administration, D.K.K.; funding acquisition, D.K.K. All authors have read and agreed to the published version of the manuscript.

**Funding:** This research was supported by Korea Institute for Advancement of Technology (KIAT) grant funded by the Korea Government (MOTIE) (RS-2025-02263945, HRD Program for Industrial Innovation). This research was supported by the Lloyd's Register Foundation (LRF, Grant No. CGY 100002). This research was a part of the project titled 'Fostering Talent in Advanced Ship Blue Tech (RS-2025-02221147)', funded by the Ministry of Oceans and Fisheries, Korea.

**Data Availability Statement:** The data are not publicly available due to commercial copyright.

**Acknowledgments:** The authors would like to express their sincere appreciation for the support provided by POSCO. Some parts of the contents are presented in the 35th International Ocean and Polar Engineering Conference (ISOPE2025), Goyang, Korea (https://onepetro.org/ISOPEIOPEC/proceedings-abstract/ISOPE25/ISOPE25/ISOPE-I-25-489/713052, accessed on 14 November 2025).

**Conflicts of Interest:** The authors declare no conflict of interest.

## Nomenclatures

The symbols used in this manuscript are defined as follows:

| Symbol | Description | Units |
|---|---|---|
| $a$ | Flaw height | mm |
| $a_0$ | Function of $V$, $L_0$ used in acceleration calculation | |
| $A$ | Function of $z$, $L_0$, $C_B$ used in acceleration calculation | |
| $a_x, a_y, a_z$ | Maximum dimensionless accelerations normalised by gravitational acceleration | |
| $B$ | Section thickness | mm |
| $B_s$ | Greatest moulded breadth of the ship | m |
| $C$ | Material constant for crack growth relationship | |
| $C_B$ | Block coefficient of the ship | |
| $c$ | Half flaw length | mm |
| $\frac{da}{dN}$ | Fatigue crack growth rate | mm/cycle |
| $F_L$ | Safety factor defined by load ratio on FAD | |
| $f_w$ | Finite width correction factor | |
| $K$ | Roll correction factor for acceleration calculation | |
| $K_c$ | Critical stress intensity factor | $\text{MPa}\sqrt{m}$ |
| $K_I$ | Stress intensity factor for tensile load (mode I) | $\text{MPa}\sqrt{m}$ |
| $K_{mat}$ | Material fracture toughness | $\text{MPa}\sqrt{m}$ |
| $K_r$ | Fracture ratio | |
| $K_{th}$ | Stress intensity factor for fatigue crack growth threshold | $\text{MPa}\sqrt{m}$ |
| $k_m$ | Stress intensity factor due to misalignment | |
| $k_{tb}$ | Bending stress concentration factor | |
| $k_{tm}$ | Membrane stress concentration factor | |
| $L_r$ | Load ratio | |
| $L_0$ | Length of the ship for scantlings | |
| $M$ | Bulging correction factor | |
| $M_b$ | Bending stress intensity magnification factor for flaws | |
| $M_m$ | Membrane stress intensity magnification factor for flaws | |
| $M_{kb}$ | Bending stress intensity magnification factor for welded joints | |
| $M_{km}$ | Membrane stress intensity magnification factor for welded joints | |
| $m$ | Exponent in flaw growth law | |
| $N$ | Strain hardening coefficient | |
| $n_d$ | Number of total cycles | cycle |
| $P_b$ | Primary bending stress | MPa |
| $P_c$ | Critical assessment point on FAD | |
| $P_i$ | Initial assessment point on FAD | |
| $P_m$ | Primary membrane stress | MPa |
| $P_0$ | Design pressure of the tank | MPa |
| $Q_b$ | Secondary bending stress | MPa |

| Symbol | Description | Units |
|---|---|---|
| $Q_m$ | Secondary membrane stress | MPa |
| $T_{cargo}$ | Design temperature of the tank | °C |
| $T_d$ | Design life in seconds | s |
| $T_{room}$ | Room temperature (typically 40 °C) | °C |
| $V$ | Service speed of the ship | knots |
| $W$ | Structure width | mm |
| $x$ | Longitudinal distance from midships to the centre of gravity of the tank | m |
| $Y$ | Stress intensity correction factor | |
| $(Y\sigma)_p$ | Primary stress intensity correction | |
| $(Y\sigma)_s$ | Secondary stress intensity correction | |
| $y$ | Transverse distance from centreline to the centre of gravity of the tank | m |
| $z$ | Vertical distance from waterline to the centre of gravity of the tank | m |
| $\alpha''$ | Parameter used in calculation of collapse stresses | |
| $\Delta K$ | Stress intensity factor range | MPa$\sqrt{m}$ |
| $\vartheta$ | Assessment angle on the FAD | ° |
| $\mu$ | Parameter for constructing FAL | |
| $\sigma_{bottom}$ | Stress at the bottom surface of the plate | MPa |
| $\sigma_{ref}$ | Reference stress | MPa |
| $\sigma_{top}$ | Stress at the top surface of the plate | MPa |
| $\sigma_u$ | Tensile strength | MPa |
| $\sigma_Y$ | Yield stress | MPa |
| $v_0$ | Average zero-crossing frequency | Hz |

## References


1. Deng, S.; Mi, Z. A review on carbon emissions of global shipping. *Mar. Dev.* **2023**, *1*, 4. [CrossRef]
2. OECD. Maritime Transport $CO_2$ Emissions. Available online: https://www.oecd.org/en/data/datasets/maritime-transport-co2-emissions.html (accessed on 20 July 2025).
3. International Maritime Organization (IMO). *2023 IMO Strategy on Reduction of GHG Emissions from Ships*; IMO Publishing: London, UK, 2023.
4. BP. Energy Outlook: 2024 Edition. 2024. Available online: https://www.bp.com/en/global/corporate/energy-economics/energy-outlook.html (accessed on 11 October 2024).
5. Al Baroudi, H.; Awoyomi, A.; Patchigolla, K.; Jonnalagadda, K.; Anthony, E.J. A review of large-scale $CO_2$ shipping and marine emissions management for carbon capture, utilisation and storage. *Appl. Energy* **2021**, *287*, 116510. [CrossRef]
6. International Maritime Organization (IMO). *International Code of the Construction and Equipment of Ships Carrying Liquefied Gases in Bulk (IGC Code)*; IMO Publishing: London, UK, 2016.
7. International Association of Classification Societies (IACS). *UR W1: Material and Welding for Ships Carrying Liquefied Gases in Bulk and Ships Using Gases or Other Low-Flashpoint Fuels*; International Association of Classification Societies: London, UK, 2021.
8. *BS 7910:2019*; Guide to Methods for Assessing the Acceptability of Flaws in Metallic Structures. British Standards Institution: London, UK, 2019.
9. EDF Energy. *R6: Assessment of the Integrity of Structures Containing Defects*; EDF Energy: Gloucester, UK, 2019.
10. *API 579-1/ASME FFS-1*; Fitness-For-Service. American Petroleum Institute: Washington, DC, USA, 2021.
11. Radu, D.; Sedmak, A.; Sedmak, S.; Li, W. Engineering critical assessment of steel shell structure elements welded joints under high cycle fatigue. *Eng. Fail. Anal.* **2020**, *114*, 104578. [CrossRef]
12. Rezaie, Y.; Sharifi, S.M.H.; Rashed, G.R. Probabilistic fracture assessment of snake laid pipelines under high pressure/high temperature conditions by engineering critical assessment. *Eng. Fract. Mech.* **2022**, *271*, 108592. [CrossRef]
13. He, M.; Pang, M.; Chen, N.Z. Engineering critical assessment (ECA) for monopile foundation of an offshore wind turbine subjected to pitting. *Ocean Eng.* **2023**, *285*, 115400. [CrossRef]
14. Seo, J.H.; Park, K.S.; Cha, I.; Choung, J. Engineering Critical Assessement for an Independent Type-B LNG Cargo Tank. *J. Soc. Nav. Archit. Korea* **2023**, *60*, 213–221. [CrossRef]

15. Kim, B.I.; Kim, K.T.; Islam, M.D.S. Development of Strength Evaluation Methodology for Independent IMO TYPE C Tank with LH2 Carriers. *J. Ocean Eng. Technol.* **2024**, *38*, 87–102. [CrossRef]
16. Kim, J.; Park, K.S.; Cha, I.; Choung, J. Structural Integrity Assessments of an IMO Type C $LCO_2$ Cargo Tank. *J. Mar. Sci.Technol.* **2025**, *13*, 1479. [CrossRef]
17. Conti, F.; Aubert, J.-M. Engineering Critical Assessment for Alternative Materials to IGC Code for $LCO_2$ Carriers Type-C Tanks. In Proceedings of the ASME 2024 43rd International Conference on Ocean, Offshore and Arctic Engineering, Singapore, 9–14 June 2024.
18. Zhen, Y.; Cao, Y.; Li, F.; Li, W.; Wu, G. A Pressure Decompression Model-Based Finite Element Approach for Efficient Dynamic Fracture Analysis in $CO_2$ Pipelines. *Fatigue Fract. Eng. Mater. Struct.* **2025**, *48*, 4245–4258. [CrossRef]
19. Anderson, T.L. *Fracture Mechanics: Fundamentals and Applications*, 4th ed.; CRC Press: Boca Raton, FL, USA, 2017.
20. Lin, Y.Y.; Jang, B.S.; Han, S.W.; Park, Y.H.; Hwang, J.; Kim, D.K. Cutting-edge insights: Crack growth rates in 9% Ni steel from room to cryogenic conditions influenced by crack closure and effective stress intensity factors. *Eng. Fract. Mech.* **2024**, *301*, 110014. [CrossRef]
21. Lin, Y.Y.; Park, K.S.; Hwang, J.; Cho, N.K.; Kim, D.K. Advanced techniques for residual stress measurement and data processing in fatigue crack growth rate tests of FH36 and 9% Ni steel at cryogenic temperatures. *Ocean Eng.* **2025**, *316*, 119942. [CrossRef]
22. *ASTM E1820-2024*; Standard Test Method for Measurement of Fracture Toughness. ASTM International: West Conshohocken, PA, USA, 2024.
23. *BS 7448-1:1991*; Fracture Mechanics Toughness Tests—Method for Determination of KIC, Critical CTOD and Critical J Values of Metallic Materials. British Standards Institution: London, UK, 1991.
24. *ISO 12135:2021*; Metallic Materials—Unified Method of Test for the Determination of Quasistatic Fracture Toughness. International Organization for Standardization: Geneva, Switzerland, 2021.
25. *ASTM E647-24*; Standard Test Method for Measurement of Fatigue Crack Growth Rates. ASTM International: West Conshohocken, PA, USA, 2024.
26. Lin, Y.Y.; Kim, Y.-J.; Cho, N.-K.; Hwang, J.H.; Park, K.-S.; Kim, D.I.; Mehmanparast, A.; Kim, D.K. Technical Recommendations for Metallic C(T) and SEN(B) Fracture Toughness Test Specimens: Data-driven Insights from Literature. *Results Eng.* **2025**, *28*, 108117. [CrossRef]
27. Abul Kashem Mohammad, Y.; Mohammad, S. Recent developments and challenges in fracture mechanics–based fatigue life prediction. *ASRC Procedia Glob. Perspect. Sci. Scholarsh.* **2025**, *1*, 1202–1237. [CrossRef]
28. Tukeev, D.L.; Afanaseva, O.V.; Tulyakov, T.F. Realization of Statistical Models Based on Symmetric Unimodal Distributions. *Int. J. Eng.* **2026**, *39*, 407–419. [CrossRef]
29. Saifi, Q. Novel probabilistic crack growth assessment method: Based on the realised PDF law for growing cracks. *Eng. Fract. Mech.* **2022**, *276*, 108931. [CrossRef]
30. Fan, X.; Chen, S.; Steingrimsson, B.; Xiong, Q.; Li, W.; Liaw, P.K. Dataset for Fracture and Impact Toughness of High-Entropy Alloys. *Sci. Data* **2023**, *10*, 37. [CrossRef] [PubMed]
31. Kim, D.K.; Ban, I.; Poh, B.Y.; Shin, S.-C. A useful guide of effective mesh-size decision in predicting the ultimate strength of flat- and curved plates in compression. *J. Ocean Eng. Sci.* **2023**, *8*, 401–417. [CrossRef]
32. Wang, Z.; Kong, W.; Wu, S.; Li, S.; Kim, D.K. A guidance of solid element application in predicting the ultimate strength of flat plates in compression. *J. Ocean Eng. Sci.* **2023**, *10*, 661–677. [CrossRef]

Journal of
*Marine Science and Engineering*

MDPI

*Article*

# Free-Drop Experimental and Simulation Study on the Ultimate Bearing Capacity of Stiffened Plates with Different Stiffnesses under Slamming Loads

**Jinsong Xia [1,2], Zhanyang Chen [3,4,*], Nan Zhao [1,2], Weidong Zhao [3,4], Qin Tang [5] and Shijian Cai [6]**

1 China Ship Scientific Research Center, Wuxi 214082, China; xjs@cssrc.com.cn (J.X.); zhaonan702@126.com (N.Z.)
2 National Key Laboratory of Ship Structural Safety, Wuxi 214082, China
3 Department of Ocean Engineering, Harbin Institute of Technology at Weihai, Weihai 264209, China; weidong.zhao@hit.edu.cn
4 State Key Laboratory of Structural Analysis, Optimization and CAE Software for Industrial Equipment, Dalian University of Technology, Dalian 116024, China
5 School of Foreign Studies, Jiangnan University, Wuxi 214122, China; 8353510219@jiangnan.edu.cn
6 Marine Design & Research Institute of China, Shanghai 200011, China; caishijian@maric.com.cn
* Correspondence: chenzhanyang@hit.edu.cn; Tel.: +86-135-0631-8766

**Abstract:** Differing from previous studies on free-drop tests, this study focuses on the ultimate bearing capacity and failure mechanism of the ship's bow under slamming loads. A prototype ship's bow is selected to design two simplified stiffened plates with different stiffeners, and the lateral slamming loads used are equivalent to flare slamming loads. Free-drop tests of the two simplified models are conducted, and the test setups and procedures are provided. The experimental results of slamming pressures and structural responses are obtained. By comparing with the simulation results obtained by Arbitrary Lagrangian-Eulerian (ALE) fluid–structure coupling, the convergence study, symmetry, and independence verifications are carried out. Finally, the dynamic ultimate bearing capacity of stiffened plates with different stiffnesses under lateral slamming loads is studied. The results show that stiffeners enhance the ability of stiffened plates to resist plastic deformation under slamming loads, and T-section stiffeners can provide greater resistance to plastic deformation than other types.

**Keywords:** free-drop model test; lateral slamming load; ALE fluid–structure coupling; ultimate bearing capacity; stiffened plate

## 1. Introduction

When a ship navigates under severe sea conditions, the slamming phenomenon resulting from the violent impact between waves and the ship's hull can cause serious structural damage, particularly to the bow [1,2]. When the ship is subjected to slamming loads, its overall bearing capacity decreases and may even be completely lost, leading to shipwrecks, which constitute a significant proportion of maritime accidents each year [1–3]. As a dynamic load, the slamming action cannot be equated to a static load due to its extremely short duration. The magnitude of the slamming load can sometimes be several times that of the static load, or even several times the wave bending moment, subjecting the ship's bow to high slamming pressures. These pressures often cause deformation or damage to the hull structure and lead to greater longitudinal bending moments in the ship. Severe impacts can not only generate immense pressure but also cause irreversible damage or complete destruction to local structures [4–6].

Compared to dynamic analysis, the static method is more commonly used in many engineering fields due to its higher computational efficiency. Thus, many studies on the ultimate strength of the ship's hull for safety evaluations have been conducted using static methods [7–10]. However, slamming is a strong nonlinear event, and it is necessary to carry out dynamic analysis for the strength of ships. Not much work or articles about experiments on the dynamic ultimate bearing capacity of structures under slamming loads have been reported. Therefore, it is essential to conduct in-depth studies on the dynamic responses and ultimate bearing capacity of the bow structure under slamming loads.

Moreover, as a highly nonlinear phenomenon, the temporal and spatial distribution characteristics of slamming pressure on the structural surface are related to many factors, such as the water entry velocity, geometric surface of the structure, elasticity of the structure, and air layers, making it challenging to fully describe using mathematical models. Thus, it is necessary to conduct experimental analysis on slamming behavior. Currently, the most commonly used methods for wave load and slamming issues are the "segmented model test" [11–14] and the "free-drop slamming test".

Compared to segmented model tests, the advantage of drop tests for slamming is that they focus more on the slamming mechanism, eliminating the interference from external factors such as waves and the ship model's navigation. Drop tests are usually conducted in towing tanks, targeting local structures. The structure impacts the still water surface in a free-fall state, and by measuring physical quantities such as slamming pressure, water entry velocity, and time of water entry, the test can predict design values or analyze the temporal and spatial distribution of slamming pressure. Based on this, many scholars have conducted numerous free-drop or controlled-speed drop tests on flat plates [15], V-shaped wedges [16], or typical wedge plate structures [17,18]. Moreover, in recent years, free-drop tests have also been conducted on the local structures of high-performance ships, such as the bow and cross-deck of trimarans. Wang et al. [19] conducted free-drop tests of two trimaran sections impacting still water to study the cross-deck slamming loads. By comparing the variation in flow field of the two models during the process of water entry, it was shown that the flow fields around the cross-decks of the two models were different. Li et al. [20] investigated the characteristics of wet deck slamming loads on a generic trimaran section through a series of free-drop slamming tests. Based on a weakening main hull method, a modified model was designed and tested to study the effect of the main hull profile on wet deck slamming. Duan et al. [21] carried out free-drop slamming tests of a trimaran section with different drop heights and heel angles to study the slamming characteristics, flow field, and trimaran motion in the process of water entry. The results showed the good symmetry and repeatability of the model test.

Most of the drop tests mentioned in the literature focus on studying the characteristics of slamming pressure, such as maximum slamming force, temporal and spatial distribution models of slamming force, and selection of slamming coefficients. However, there is still limited research on the plastic deformation and ultimate bearing capacity of structures under slamming loads due to limitations in experimental conditions and experimental cost.

Therefore, to achieve free-drop tests for the bearing capacity of structures, a certain ship's bow is simplified into two sets of wedge grillage with a water entry angle of 15°, thereby achieving the equivalence of flare slamming with lateral slamming loads. The rest of the paper is organized as follows. In Section 2, the test system, test model, data acquisition system, and test program are described in detail. The ALE fluid–structure coupling simulation is conducted in Section 3, including the model establishment, parameter setting and data presentation. In Section 4, a verification study of the repeatability, linear regression and independence of the different model for the test is conducted. The main results of the free-drop tests are presented in Section 5, focusing on the plastic deformation and ultimate bearing capacity. Moreover, the influence of different models on the failure mechanism is discussed. Finally, conclusions are drawn, and an outlook for future work is presented in Section 6.

## 2. Experimental Setup

The primary objective of this experiment is to conduct model tests on the dynamic ultimate bearing capacity of stiffened plates with different stiffness levels under lateral slamming loads using a free-drop test approach. By measuring the slamming pressure, structural strain, and plastic deformation of the plates, the load characteristics and failure mechanisms of the stiffened plates are determined under lateral slamming loads.

## 2.1. Model Design

Considering the magnitude of the dynamic ultimate strength of stiffened plates and the experimental conditions, the grillage structure of a ship's bow is simplified by ignoring the curvature of the outer plate, and the four simplified stiffened plate models with different specifications are designed, including two non-stiffened plates, one stiffened plate with angled steel, and one stiffened plate with a T-section stiffener. The geometric parameters are shown in Table 1. #1 and #2 are non-stiffened plates with different thicknesses, and #3 and #4 are stiffened plates with different longitudinal stiffeners. Considering that a small water entry angle might cause cushion effects, the water entry angle for all experimental models is set to 15°. The material for the models is ordinary steel (Q235), which is widely used in many industrial fields. The density is 7850 kg/m$^3$, the elastic modulus is $2.1 \times 10^{11}$ Pa, the yield strength is 235 MPa and the Poisson's ratio is 0.3. The steel sheets selected are hot-rolled.

The primary purpose of this experiment is to study the ultimate bearing capacity of stiffened plates under lateral slamming loads, which is a destructive test. The experimental models and sectional view are shown in Figure 1. Model 1 is comprises #1 and #3, and Model 2 comprises #2 and # 4. The total height of the stiffened plate and box structure models is 1000 mm, and the self-weight of the experimental model is approximately 750 kg (excluding ballast). The average draft of each model is around 0.5 m, and the height of the center of gravity is around 0.7 m. To prevent splashes from entering the box after water entry, a 0.4 m high splash guard is added to the top of the box during fabrication, which also enhances the sinking resistance of the entire experimental model.

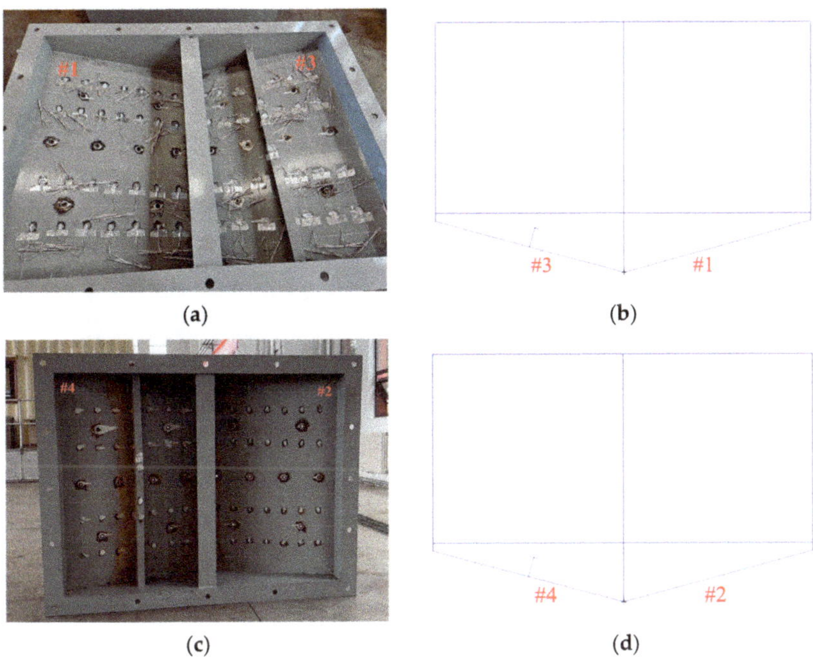


**Figure 1.** View of model: (**a**) Model 1 for test; (**b**) Diagram of Model 1; (**c**) Model 2 for test; (**d**) Diagram of Model 2.

**Table 1.** Structural parameters of stiffened plates.

| No. | Length (mm) | Width (mm) | Thickness (mm) | Number of Stiffeners | Height of Web (mm) | Thickness of Web (mm) | Width of Plate (mm) | Thickness of Plate (mm) |
|---|---|---|---|---|---|---|---|---|
| #1 (Non-stiffened) | 1200 | 800 | 2 | 0 | -- | -- | -- | -- |
| #2 (Non-stiffened) | 1200 | 800 | 3 | 0 | -- | -- | -- | -- |
| #3 (Angled steel) | 1200 | 800 | 2 | 1 | 100 | 2 | 20 | 3 |
| #4 (T-section stiffener) | 1200 | 800 | 2 | 1 | 100 | 2 | 40 | 3 |

### *2.2. Instruments*

The experiment was conducted in the towing tank at the Qingdao Branch of China Ship Scientific Research Center, with a total depth of 10–12 m and an adjustable water surface height. The lifting device is 9 m above the ground. To ensure that the model falls in a normal state (with the bottom of the model level), the lifting device is used several times before the experiment to find the model's center of gravity, and the center of gravity is adjusted to the mid-longitudinal plane through weighting. During the experiment, the model is hoisted by the carriage, with a release device between the hook and the steel cable. To stabilize the model during the test, two traction ropes are attached to both ends longitudinally to adjust the model's position. In addition, a rope is attached to the lifting shackle, with its length adjusted to ensure that the steel mold does not accidentally sink and to facilitate lifting the steel cable after the model enters the water. The release process is completed using a wireless remote control to open the release device. The suspension setup for the experimental model is shown in Figure 2.

**Figure 2.** Suspension device for the experimental model.

In this study, the measured data acquired from the experimental procedure mainly include the slamming pressure and structural strain, which are measured by the pressure sensors and strain sensors, respectively. The pressure sensors used in the experiment have a range of 500 kPa to 1 MPa, and the accelerometers have a range of 5 g. The strain sensors are unidirectional strain gauges. Data acquisition systems are used to collect measurement data. All instruments and sensors are in good condition and within their calibration validity period. The wiring and instrument layout of the experimental model are shown in Figure 3a,b. Before and after the test, structural deformation measurements of the experimental model are taken using the scanner, as shown in Figure 3c.

(a) (b) (c)

**Figure 3.** Main test equipment: (**a**) Wiring of experimental model; (**b**) Data acquisition instruments; (**c**) Structural deformation scanner.

### 2.3. Test Conditions and Content

It is well known that the slamming pressure depends on the water entry velocity of the structure [22,23]. This experiment ensures that the model reaches the rated water entry velocity by adjusting the drop height. The drop height is defined as the distance from the lowest point of the model to the water surface. The model is lifted by the lifting device, and the height of the model from the ground can be measured by using a laser range finder at the edge of the towing tank. Since the distance between the still water and the ground is fixed, the distance between the model and the still water, which is the drop height, can be obtained. The experiment includes five different drop heights: 1.0 m, 1.5 m, 2.5 m, 5.1 m, and 7.34 m. The drop heights at which the two models respectively reached the plastic deformation are 5.1 m and 7.34 m. The specific conditions are presented in Table 2.

**Table 2.** Simulation conditions.

| Model | Case | Water Entry Angle (deg) | Drop Height (m) | Water Entry Velocity (m/s) |
|---|---|---|---|---|
| 1 | 1 | 15 | 1.00 | 4.43 |
| | 2 | 15 | 1.50 | 5.42 |
| | 3 | 15 | 2.50 | 7.00 |
| | 4 (Structure failure) | 15 | 5.10 | 10.00 |
| 2 | 5 | 15 | 5.10 | 10.00 |
| | 6 (Structure failure) | 15 | 7.34 | 12.00 |

## 3. Numerical Simulation

In this study, the ALE fluid–structure coupling method available in commercial LS-DYNA software is used to study the water entry of elastic stiffened plates [19,20]. To compare the simulation results with the experimental results, considering the high-frequency characteristics of the slamming pressure, the sampling frequency for the simulation is 2 kHz. Additionally, to save computational resources, the model's drop height is shortened in the simulation, only ensuring that the model meets the water entry velocity.

### 3.1. Simulation Model and Parameters

The simulation model of the upper box structure is shown in Figure 4a, which is set as a rigid body to reduce the simulation time. The stiffened plate models, which are made of elasto-plastic materials, are presented in Figure 4b,c. In this study, the wedge grillage is a thin shell structure modeled using the quadrilateral mesh. After verification of grid independence (as shown in Section 4.1), the mesh size of each wedge grillage is determined to be 20 mm × 20 mm. The total mesh number of the wedge grillage is 33,312.

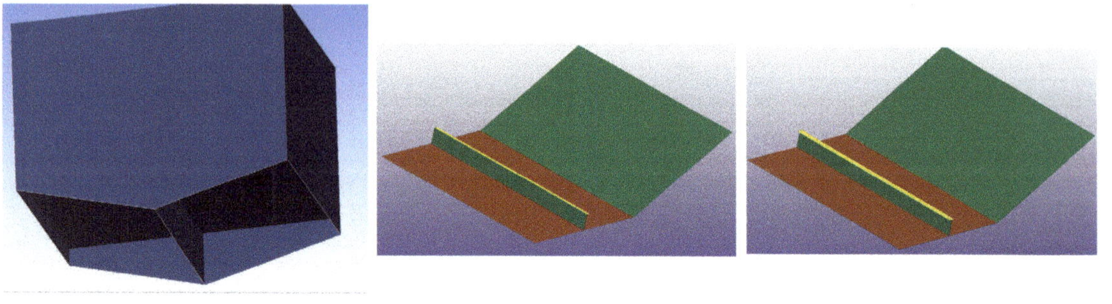

(**a**) (**b**) (**c**)

**Figure 4.** Simulation structural model: (**a**) Upper box model; (**b**) Angled steel stiffened plate (#1 and #3); (**c**) T-bar stiffened plate (#2 and #4).

In addition to the wedge model, the ALE method also requires constructing a fluid domain model, including the air domain and the water domain. The dimensions of the water domain are 4.5 m × 2.4 m × 1.25 m, and the air domain dimensions are also 4.5 m × 2.4 m × 1.25 m. To ensure computational accuracy and reduce the computation time, the fluid domain model is meshed with a gradient grid, with refined grids in the middle and coarse grids around the edges. The total mesh number of fluid domain is 883,200.

Furthermore, during the simulation, the reference atmospheric pressure is applied at the boundaries of the fluid domain rather than the entire fluid. The gravity acceleration of 9.81 kN/m$^2$ is applied vertically downward to all boundary layers, along with a reference atmospheric pressure of 1.013 × 10$^5$ Pa. The water entry of the simulation model is presented in Figure 5.

**Figure 5.** Water entry of simulation model.

*3.2. Monitoring Points and Data Presentation*

The simulation study corresponds with the experiment, and the arrangement of monitoring points are presented in Figure 6. Each stiffened plate has 8 pressure monitoring points (denoted by P) and 32 strain monitoring points (denoted by S). S15/16 and S17/18 indicate one strain gauge on the web and one on the panel of the stiffener, respectively.

(a)

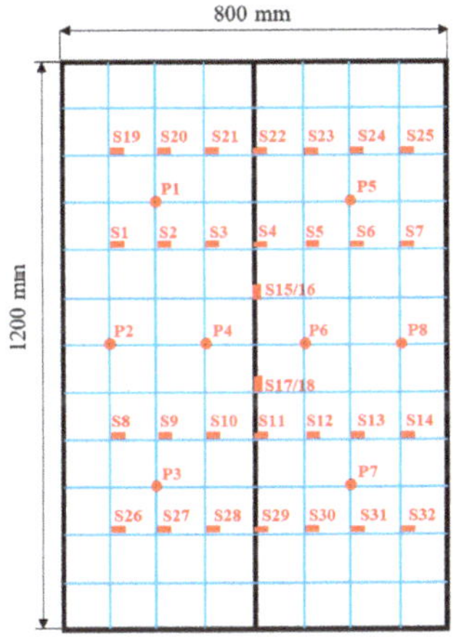

(b)

**Figure 6.** Arrangement of pressure monitoring points of Model 2: (**a**) Test model; (**b**) Pressure and strain monitoring points for #4.

The time series of slamming pressure of Model 1 in Case 4 are presented in Figure 7. It can be seen that slamming occurs around 0.1 s, and the higher the entry point, the lower the water entry velocity and pressure. The pressure values at the same vertical height are almost equal.

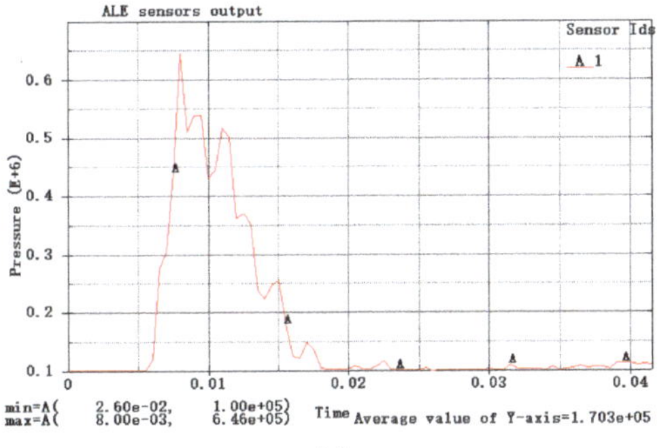

(a)

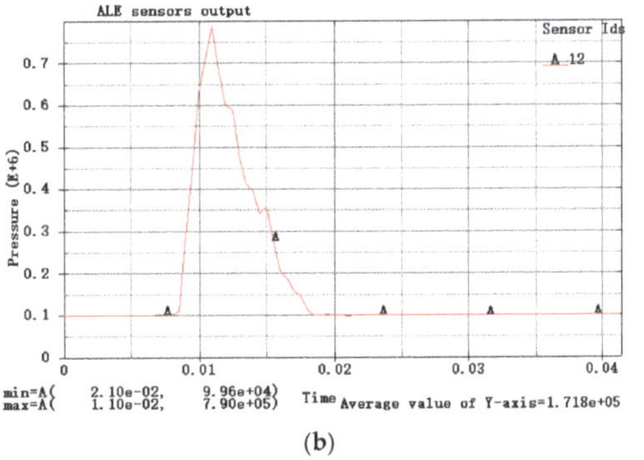

(b)

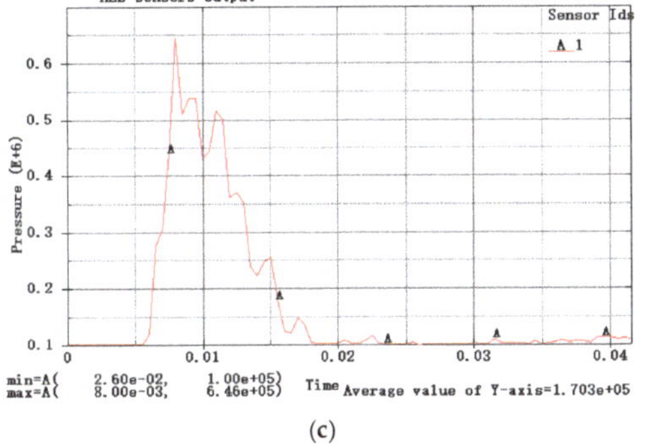

(c)

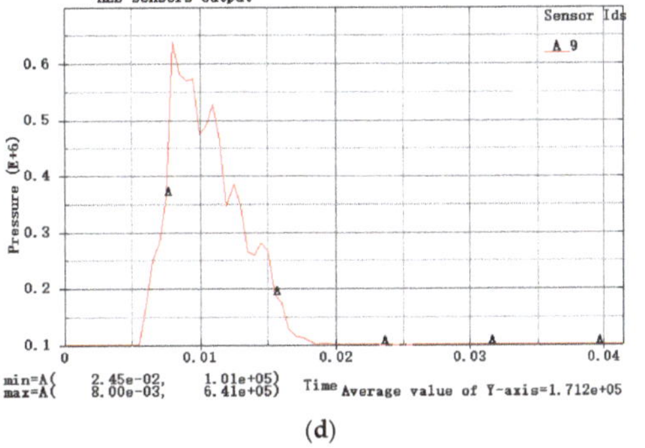

(d)

**Figure 7.** *Cont.*

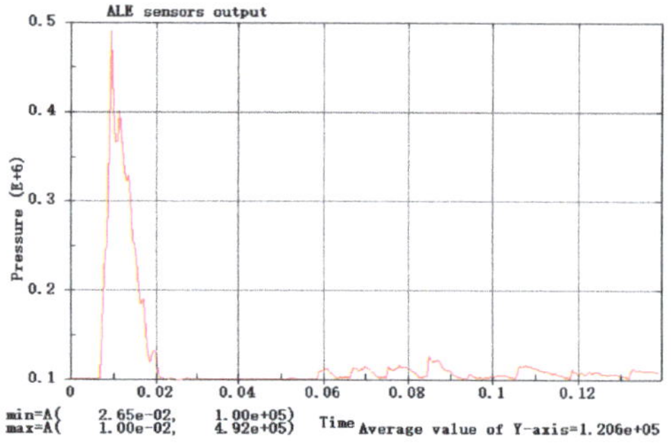


(e)

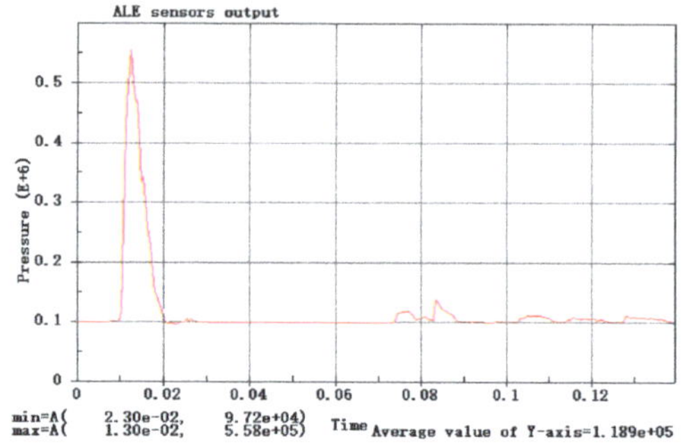


(f)

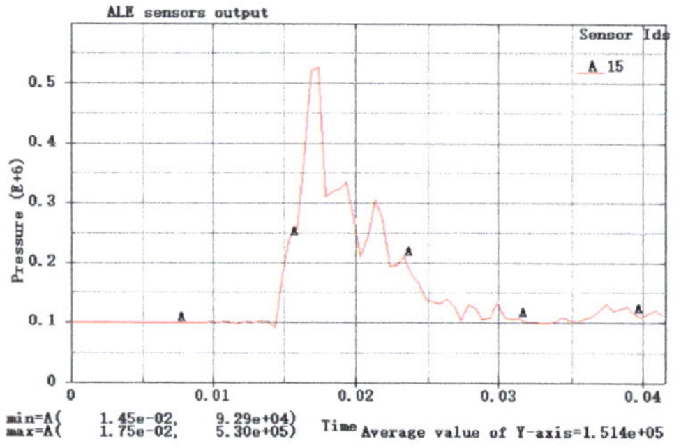


(g)

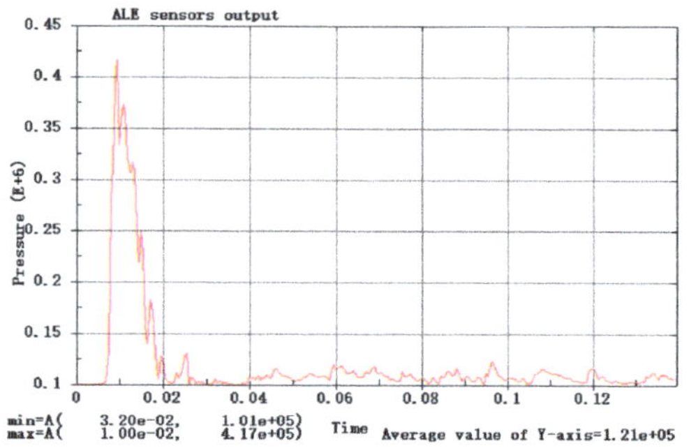


(h)

**Figure 7.** Time series of simulated slamming pressure in Case 4 of Model 1: (**a**) P1; (**b**) P2; (**c**) P3; (**d**) P4; (**e**) P5; (**f**) P6; (**g**) P7; (**h**) P8.

## 4. Comparison and Verification

This study mainly focuses on the plastic deformation and failure mechanism of structures. As the input for system response, the slamming pressure is the first to be measured. Therefore, the measured slamming pressure is selected to verify the reliability of the numerical and experimental results.

### *4.1. Convergence Study*

To further verify the numerical simulation used in this work, it was necessary to conduct a convergence study. According to the methodology employed by Stern and Wilson [24], the convergence study of the grid is carried out in this section. Model 2 and Case 6 are selected, and the verification focuses on the slamming pressure of P2. In this study, four grid schemes, which are 80 mm × 80 mm, 40 mm × 40 mm, 20 mm × 20 mm, and 10 mm × 10 mm, respectively, were designed for the verification study.

The time series of the slamming pressure of P2 for the four grid systems were simulated and are shown in Figure 8. As can be seen, the overall trend of the slamming pressures obtained based on four grid schemes is similar, but the peak values differ significantly. The slamming pressure corresponding to the grid scheme of 80 mm × 80 mm is the lowest, and the slamming pressure increases with refinement of the mesh. The slamming pressure corresponding to the grid scheme of 20 mm × 20 mm is significantly higher than that of 40 mm × 40 mm, and the timing of peak occurrence is also different. In addition, when the mesh size is refined to 10 mm × 10 mm, the slamming pressure slightly increases, but the computation time significantly increases. Thus, to achieve a compromise between computation efficiency and accuracy, the grid scheme of 20 mm × 20 mm was selected for the subsequent simulation.

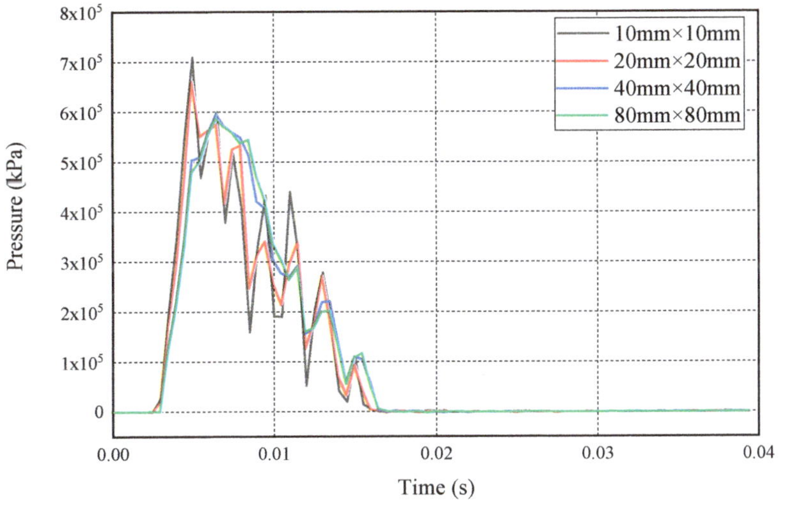


**Figure 8.** Time series of slamming pressure of P2 in Case 6 of Model 2.

### 4.2. Symmetry Verification

The time series of the slamming pressures at monitoring points P1 and P5 for Case 5 of Model 2 were simulated and are presented in Figure 9. It can be seen from Figure 6 that the same numbered points on #2 and #4 are symmetrically arranged. They enter the water simultaneously at the same height during the experiment, with P1 entering the water before P5.

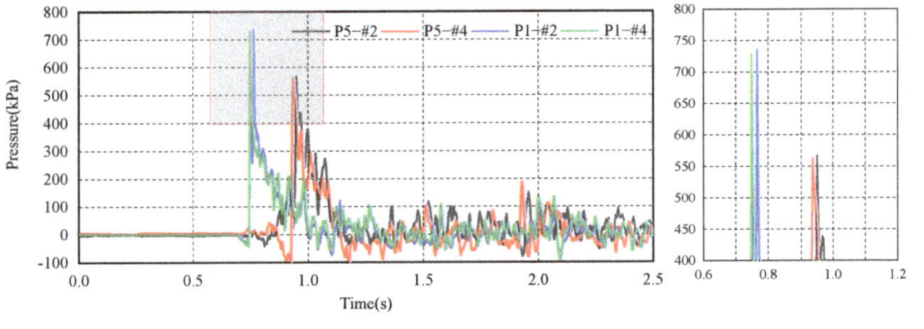


**Figure 9.** Time series of measured slamming pressure in Case 5 of Model 2.

As shown in Figure 9, the slamming pressure's time distribution resembles a pulse load pattern, with the pressure reaching its maximum at the moment of water entry and then decaying to zero. Since the pressure peak is proportional to the square of the water entry velocity, the same height results in the same water entry velocity, leading to nearly equal slamming pressure peaks at the same symmetric positions (P1-#2 and P1-#4; P5-#2 and P5-#4). This indicates that the physical process of slamming and the geometric shape of the model have good symmetry. The earlier water entry point reaches the pressure peak first. The earlier entry point has a higher pressure peak because it starts decelerating upon contact with the water. Furthermore, the entry point might have multiple peaks due to water splashing. The duration of the first peak is very short, which is followed by high-frequency oscillations. The times when the first peak appears, corresponding to the same height, are almost the same. The second peak of the time series has obvious randomness.

### 4.3. Independence of the Different Models

In this section, to verify the independence of the different test models, the experimental and simulation results of the slamming pressures of four stiffened plates at the same water entry velocity (10 m/s) are compared and presented in Figure 10. It can be seen that the test results for different stiffened plates at the same water entry velocity are not significantly different. This is because the slamming pressure peak only depends on the water entry velocity. Moreover, compared to the simulation results of different stiffened plates, the test results are slightly higher. The quantification of the deviation is presented in Table 3. As can be seen, deviations 1–4 denote the differences between the measured pressures of the four stiffened plates (#1−#4) and the simulation results, respectively. The maximum deviation occurs at P4 of stiffened plate #3, which is 10.36%, and the minimum deviation occurs at P7 of stiffened plate #4, which is 0.33%. It is found that the amount of deviation is relatively stable, proving the stability of the experimental measurements. Considering minor differences in instrumentation operation during each test, the experimental results are reliable.

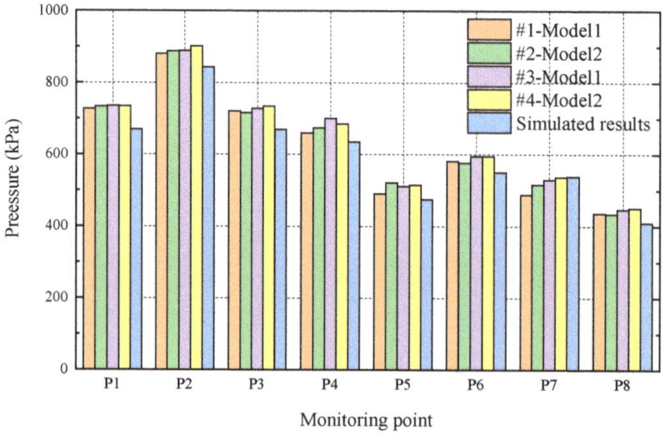


**Figure 10.** Comparison between measured pressure and simulated pressure.

**Table 3.** Quantification of the deviation in Figure 10 (kPa).

| Point | #1-Model 1 | #2-Model 2 | #3-Model 1 | #4-Model 2 | Simulations | Deviation 1 | Deviation 2 | Deviation 3 | Deviation 4 |
|---|---|---|---|---|---|---|---|---|---|
| P1 | 726.28 | 732.76 | 734.53 | 733.76 | 669.00 | 8.56% | 9.53% | 9.79% | 9.68% |
| P2 | 880.72 | 888.01 | 889.00 | 901.84 | 844.00 | 4.35% | 5.21% | 5.33% | 6.85% |
| P3 | 720.69 | 716.19 | 727.97 | 734.07 | 669.00 | 7.73% | 7.05% | 8.82% | 9.73% |
| P4 | 659.85 | 673.92 | 700.80 | 685.21 | 635.00 | 3.91% | 6.13% | 10.36% | 7.91% |
| P5 | 490.66 | 521.66 | 511.55 | 515.64 | 475.00 | 3.30% | 9.82% | 7.69% | 8.56% |
| P6 | 581.92 | 576.74 | 595.45 | 595.25 | 551.00 | 5.61% | 4.67% | 8.07% | 8.03% |
| P7 | 487.62 | 516.41 | 529.82 | 537.20 | 539.00 | 9.53% | 4.19% | 1.70% | 0.33% |
| P8 | 436.65 | 434.39 | 446.93 | 450.70 | 410.00 | 6.50% | 5.95% | 9.01% | 9.93% |

## 5. Results and Discussion

### 5.1. Distribution Pattern of Structural Strain

To meet the requirements of measurement accuracy, the microstrain was measured as the direct output by the strain gauges during the experiment. The time series of microstrain on the stiffened plates under Case-4 of Model 1 are presented in Figure 11. It is evident that the microstrain distribution over time is synchronized with the slamming pressure, occurring around 0.75 s. The plastic deformation can be seen from the time series of microstrain. The spatial distribution of microstrain shows that the plastic microstrain on the stiffened panel is the highest, with the plastic microstrain reaching 8000 at point S16 and reaching 200,000 at point S18. The stiffened web takes second place, with the plastic microstrain ranging from 3000 to 6000, and the plastic microstrain of the plate grids being below 4000.

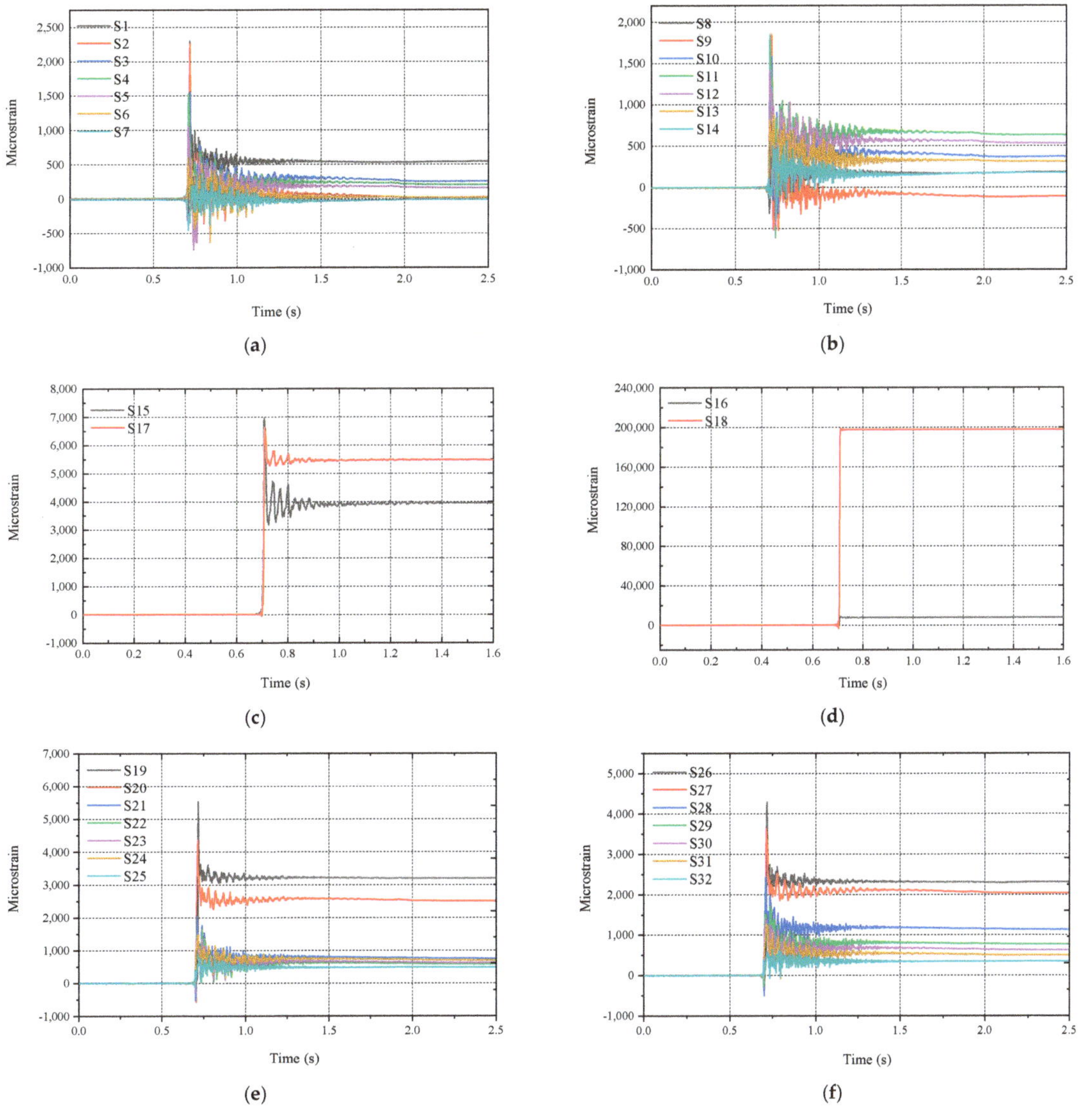


**Figure 11.** Time series of microstrain in Case 4: (**a**) S1–S7; (**b**) S8–S14; (**c**) S19–S25; (**d**) S26–S32; (**e**) S15 and S17 for web; (**f**) S16 and S18 for plate.

### *5.2. Analysis of Plastic Deformation of Stiffened Plates under Lateral Slamming Loads*

From the time series in Figure 10, it is evident that the test models undergo plastic deformation under slamming loads. It is noted that the two models in each case can undergo varying degrees of plastic deformation, but the cases under which full plastic deformation occurs differ between the two stiffened plate models. The stiffened plate with angled steel (Model 1) undergoes plastic deformation at the water entry velocity of 10 m/s (Case 4), while the stiffened plate with a T-section stiffener (Model 2) undergoes plastic deformation at the water entry velocity of 12 m/s (Case 6). The post-test deformations of the stiffened plates are shown in Figure 12.

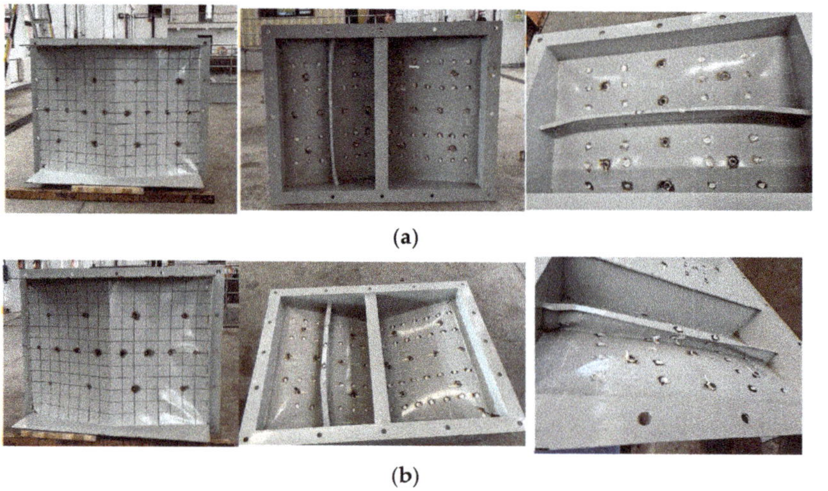

**Figure 12.** Plastic deformation of the two models: (**a**) Model 1 (Case 4); (**b**) Model 2 (Case 6).

As shown in Figure 12, both models exhibit extensive plastic deformation, with damage to the plate grids, stiffened panels, and stiffened webs. This indicates that the dynamic ultimate strength failure test of stiffened plates under lateral slamming loads is completed in the study. As shown in Figure 12a, the plastic deformation in the middle of the plate grid and along the diagonal is significant, leading to a half-inverted roof shape. As shown in Figure 12b, the failure mode of the plate grid of Model 2 is the formation of plastic hinges along the diagonal of the plate edge, with concave deformation occurring in the middle of the plate grid. The failure mode of the stiffener is the formation of plastic hinges at the ends (boundaries) of both the web and panel, causing lateral deformation of the stiffener.

To more intuitively observe the deformation of the stiffened plates after testing, scanning measurements were conducted on the stiffened plate components under various conditions. The structural deformation measurements of each component after the test are presented in Figures 13 and 14. As shown in Figures 13 and 14, the deformation of the non-stiffened plate exhibits a symmetrical shape, while the stiffened plate exhibits an asymmetrical shape, with the deformation of the plate being greater than that of the stiffened plate, indicating that stiffening enhances the resistance to deformation. Moreover, it is noted that although the maximum deformations of the stiffened plate and stiffener in Case 6 are slightly larger than those in Case 4, the water entry velocity of Model 2 with full plastic deformation is greater. Thus, the T-section stiffener has stronger resistance to deformation under slamming loads.

Finally, the deformation trends of different components in each model after testing are presented in Figure 15. It can be seen that the overall deformation for different components in each model increases with increasing water entry velocity (drop height). Furthermore, the plate grid deformation and stiffener deformation of the T-section stiffened plate are smaller than those of the angled steel stiffened plate under the same water entry velocity, indicating that the T-section stiffener has a stronger ability to resist plastic deformation under slamming loads.

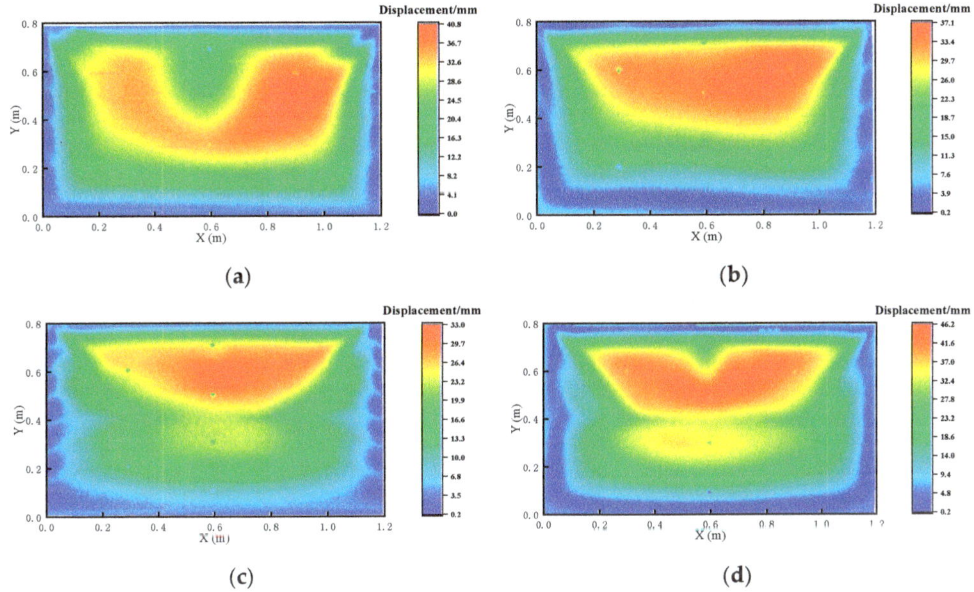


**Figure 13.** Deformation of plates: (**a**) #1 (Case 4); (**b**)#2 (Case 6); (**c**) #3 (Case 4); (**d**) #4 (Case 6).

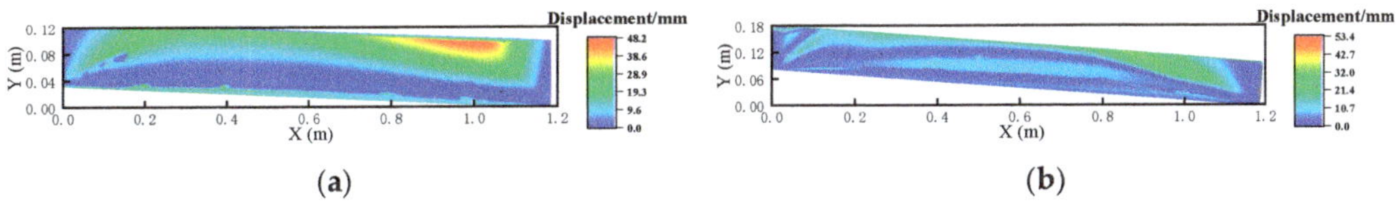


**Figure 14.** Deformation of stiffeners: (**a**) Angled steel; (**b**) T-section steel.

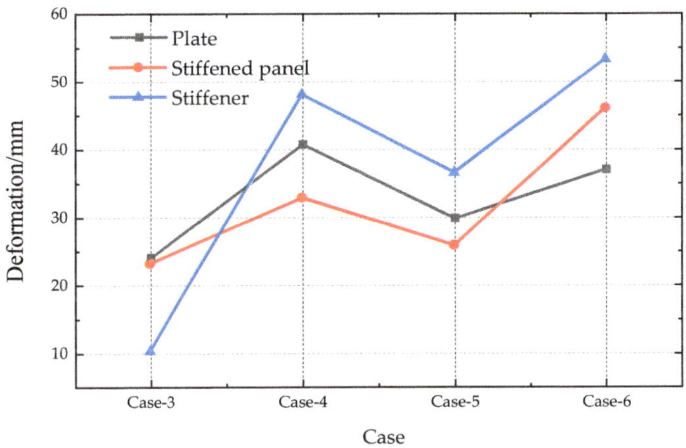


**Figure 15.** Deformation trend of different members after each model test.

## 6. Conclusions

By using numerical simulations based on LS-DYNA software (LS-DYNA_mpp_s_R11_0) and free-drop slamming tests, the dynamic ultimate bearing capacity of the structure under lateral slamming loads, including structural strain and plastic deformation, was analyzed for stiffened plates with different stiffnesses. The focus of the study was on analyzing the influence of factors such as different stiffness, position, and water entry velocity on the plastic deformation of the structure. The following conclusions can be drawn:

(1) The time distribution pattern of slamming pressure resembles a pulse load, with the pressure reaching its maximum at the moment of water entry. The spatial distribution of slamming pressure is related to the order of water entry. Points entering the water simultaneously at the same vertical height have similar peaks of slamming pressure, with earlier entry points experiencing higher peak pressures. Points entering later may have multiple peaks due to splashing water.

(2) The time series of microstrain is synchronized with the slamming pressure, and plastic deformation occurs in the structure. The spatial distribution shows that the plastic deformation on the stiffened panel is the highest.

(3) Under slamming loads, the failure mode of the plate and stiffened plate grids involves the formation of plastic hinges along the plate edge diagonals, with concave deformation occurring in the middle of the plate grids. The failure mode of the stiffener involves the formation of plastic hinges at the ends (boundaries) first, followed by stiffener lateral deformation.

(4) Stiffeners can enhance the ability of stiffened plates to resist plastic deformation under slamming loads, with T-section stiffeners providing stronger resistance to plastic deformation. Increasing the plate thickness can improve the ability to resist plastic deformation.

It is noted that the water entry angle of 15° was selected in this study based on past experimental experience, and the influence of the water entry angle on the slamming pressure and structural response characteristics needs be determined. Thus, free-drop experiments of models with different water entry angles will be conducted in the future.

**Author Contributions:** Conceptualization, J.X. and Z.C.; methodology, J.X.; software, Z.C. and W.Z.; validation, J.X., Z.C. and W.Z.; formal analysis, S.C.; investigation, Z.C.; resources, J.X.; data curation, N.Z. and S.C.; writing—original draft preparation, Z.C. and Q.T.; writing—review and editing, Z.C. and Q.T.; visualization, W.Z.; supervision, J.X. and N.Z.; project administration, J.X. and S.C.; funding acquisition, J.X. and Z.C. All authors have read and agreed to the published version of the manuscript.

**Funding:** This research was funded by the State Key Laboratory of Structural Analysis, Optimization and CAE Software for Industrial Equipment, Dalian University of Technology (Grant number GZ23112 and GZ23109). The APC was funded by the China Ship Scientific Research Center, Wuxi.

**Institutional Review Board Statement:** Not applicable.

**Informed Consent Statement:** Not applicable.

**Data Availability Statement:** All data, models, or code generated or used during the study are available from the corresponding author by request.

**Conflicts of Interest:** The authors declared that they have no conflicts of interest to this work.

## References

1. AbuBakar, A.; Dow, R.S. Simulation of ship grounding damage using the finite element method. *Int. J. Solids Struct.* **2013**, *50*, 623–636. [CrossRef]
2. Ehlers, S. The influence of the material relation on the accuracy of collision simulations. *Mar. Struct.* **2010**, *23*, 462–474. [CrossRef]
3. Liu, B.; Pedersen, P.T.; Zhu, L.; Zhang, S. Review of experiments and calculation procedures for ship collision and grounding damage. *Mar. Struct.* **2018**, *59*, 105–121. [CrossRef]
4. Howison, S.D.; Ockendon, J.R.; Wilson, S.K. Incompressible water entry problems at small deadrise angles. *J. Fluid Mech.* **1991**, *222*, 215–230. [CrossRef]
5. Stenius, I.; Rosén, A.; Kuttenkeuler, J. Explicit FE-Modelling of Fluid-Structure Interaction in Hull-Water Impacts. *Int. Shipbuild. Prog.* **2006**, *53*, 103–121.
6. Ren, H.; Yu, P.; Wang, Q.; Li, H. Dynamic response of the bow flare structure under slamming loads. In Proceedings of the 34th International Conference on Ocean, Offshore and Arctic Engineering (ASME 2015), St. John's, NL, Canada, 31 May–5 June 2015.

7. Iijima, K.; Fujikubo, M. Hydro-elastoplastic behaviour of VLFS under extreme vertical bending moment by segmented beam approach. *Mar. Struct.* **2018**, *57*, 1–17. [CrossRef]
8. Yang, L.; Peng, Z.L.; Wang, D.Y. Experimental and numerical investigation of material failure criterion with high-strength hull steel under biaxial stress. *Ocean Eng.* **2018**, *155*, 24–41. [CrossRef]
9. Yang, B.; Soares, C.G.; Wang, D.Y. Dynamic ultimate compressive strength of simply supported rectangular plates under impact loading. *Mar. Struct.* **2019**, *66*, 258–271. [CrossRef]
10. Yang, B.; Wang, D. Buckling strength of rectangular plates with elastically restrained edges subjected to in-plane impact loading. *Proc. Inst. Mech. Eng. Part C J. Mech. Eng. Sci.* **2017**, *231*, 3743–3752. [CrossRef]
11. Zhu, S.J.; Wu, M.K.; Moan, T. Experimental investigation of hull girder vibrations of a flexible backbone model in bending and torsion. *Appl. Ocean Res.* **2011**, *33*, 252–274. [CrossRef]
12. Stenius, I.; Rosén, A.; Battley, M.; Allen, T. Experimental hydroelastic characterization of slamming loaded marine panels. *Ocean Eng.* **2013**, *74*, 1–15. [CrossRef]
13. Chen, Z.Y.; Jiao, J.L.; Li, H. Time-domain numerical and segmented ship model experimental analyses of hydroelastic responses of a large container ship in oblique regular waves. *Appl. Ocean Res.* **2017**, *67*, 78–93. [CrossRef]
14. Chen, Z.Y.; Gui, H.B. Research on Whipping and Springing Responses of Hull based on Different Experimental Method and Nonlinear Hydroelastic Time-domain Theory. *J. Ship Mech.* **2017**, *21*, 1143–1158.
15. Luo, H.; Liu, X.; Dong, D.; Zhao, Z.; Ji, H.; Gao, Y.; Lin, W. Experimental investigation of water impact of one free-drop wedge with stiffened aluminum panels. *Chin. J. Hydrodyn.* **2014**, *29*, 460–468.
16. Si, H.L.; Zhao, N.; Hu, J.J. Experimental study of slamming loads on ship bow during free water entry. *J. Ship Mech.* **2020**, *24*, 445–455.
17. Mo, L.X.; Wang, H.; Jiang, C.X.; Xu, C. Study on dropping test of wedge grillages with various types of stiffeness. *J. Ship Mech.* **2011**, *15*, 394–401.
18. Chen, X.P.; Li, J.W.; Wang, H.; Qi, E.R. Experiments and numerical investigation of water entry of large scale steel wedge models. *J. Ship Mech.* **2012**, *16*, 1152–1163.
19. Wang, Q.; Yu, P.Y.; Fan, G.J.; Li, G.Z. Experimental drop test investigation into cross deck slamming loads on a trimaran. *Ocean Eng.* **2021**, *240*, 109999. [CrossRef]
20. Li, H.; Deng, B.L.; Zou, J.; Dong, C.R.; Liu, C.L.; Liu, P.L. Experimental free-drop test investigation into wet-deck slamming loads on a generic trimaran section considering the influence of main hull profile. *Ocean Eng.* **2021**, *242*, 110114. [CrossRef]
21. Duan, W.Y.; Liu, J.Y.; Liao, K.P.; Ma, S. Experimental study of slamming pressure for a trimaran section with different drop heights and heel angles. *Ocean Eng.* **2022**, *263*, 112400. [CrossRef]
22. Stavovy, A.B.; Chuang, S.L. Analititcal determination of slamming pressure for hign speed vehicals in waves. *J. Ship Res.* **1976**, *20*, 190–198. [CrossRef]
23. Ochi, M.D.; Bonilla-Norat, J. *Pressure-Velocity Relationship in Impact of a Ship Model Dropped onto the Water Surface and in Slamming in Waves*; AD-709071; National Academy of Sciences: Washington, DC, USA, 1970.
24. Stern, F.; Wilson, R.V.; Coleman, H.W.; Paterson, E.G. Comprehensive approach to verification and validation of CFD simulations—Part 1: Methodology and procedures. *J. Fluids Eng.* **2001**, *123*, 793–802. [CrossRef]

Journal of
*Marine Science and Engineering*

MDPI

*Article*

# A Comparative Study on the Calculation Methods of Nonlinear Springing of Large Containerships

**Ye Lu [1,2], Liye Li [1,2] and Qiu Jin [3,4,*]**

1 China Ship Scientific Research Center, Wuxi 214082, China; luye@cssrc.com.cn (Y.L.); 4_panda@163.com (L.L.)
2 Taihu Laboratory of Deepsea Technological Science, Wuxi 214082, China
3 Key Laboratory of High Performance Ship Technology (Wuhan University of Technology), Ministry of Education, Wuhan 430063, China
4 School of Naval Architecture, Ocean and Energy Power Engineering, Wuhan University of Technology, Wuhan 430063, China
* Correspondence: autumnjinqiu@whut.edu.cn

**Abstract:** The vibration of large containerships induced by waves and its resulting fatigue damage have long been the focus of research in the field of ocean engineering. For low-frequency nonlinear wave-induced springing, potential flow and viscous flow remain the two significant calculation methods. Based on potential flow theory, this study investigates the nonlinear wave-induced vibration response of large containerships, including the superposition of sum and difference frequencies, by considering the influence of second-order hydrodynamic forces. Meanwhile, a three-dimensional numerical wave basin model is established to simulate fluid–structure interaction, integrating structural mode superposition for two-way CFD (Computational Fluid Dynamics)-FEM (Finite Element Method) coupling. By comparing with the experimental results, it is found that the frequency-domain nonlinear method considering second-order hydrodynamic forces and the CFD-FEM method can both effectively capture the nonlinear wave-induced vibration phenomenon under regular wave conditions. The numerical simulation results of the two methods are close to the experimental results. Moreover, the frequency-domain nonlinear method has a fast calculation speed, making it more suitable for the preliminary design of large ships.

**Keywords:** large containership; nonlinear wave-induced springing; nonlinear hydroelasticity in frequency domain; CFD-FEM coupled hydroelasticity

## 1. Introduction

As globalization and the integration of the world economy continue to advance, the demand for large containers has surged, leading to a growing preference among carriers for ultra-Panamax containerships and even larger vessels. This plays a significant role in load valuation for use in the design of large ships [1]. Large containerships are prone to linear wave-induced vibrations at high speeds, and even after slowing down, they still experience intense nonlinear wave-induced vibrations. The impact of these nonlinear wave-induced vibrations on the fatigue strength of ship structures cannot be ignored [2].

Currently, in the field of ship and marine engineering, the most widely used theory for evaluating the motion of floating bodies in waves is three-dimensional potential flow theory. Generally, this theory assumes that the floating body is a rigid body. However, most actual floating bodies are steel structures, which are elastic. Wu [3] and Price [4] innovatively combined three-dimensional hydrodynamic theory with three-dimensional structural dynamics theory, proposing generalized fluid–structure boundary conditions. Based on three-dimensional potential flow theory, they assumed homogeneous, inviscid, irrotational, and incompressible fluids, with their free surfaces being small-amplitude waves. The linearly elastic floating structure has minimal motion and vibration deformation relative to its equilibrium position. Using modal superposition methods, they studied the interaction phenomena between inertial forces, hydrodynamic forces, and elastic forces, unifying the study of the effects of the fluid on the structure and the structure's response. This led to the development of a three-dimensional linear hydroelastic mechanics theory applicable to any three-dimensional deformable body moving in waves or navigating underwater, analyzing the structural dynamic responses under internal and external excitation.

Under high sea conditions, when ships experience significant motion, the primary distinction from the assumptions of linear hydroelastic theory is that the ship's movements are not minor but involve large-angle rigid body rotations. Concurrently, the free surface of the surrounding fluid is no longer an average wetting surface (or the wetting surface during static floating). In such scenarios, second-order forces in hydrodynamics frequently cannot be overlooked, and the corresponding nonlinear structural responses are also considerable. Wu et al. [5] addressed the nonlinear situations caused by large-angle rigid body rotations and instantaneous wet surfaces. For six-degree-of-freedom rigid body motions of floating structures in rough seas, their angles must be expressed using the first two terms of the appropriate Taylor series expansion. The instantaneous wet surface of the floating structure cannot be simplified as an average wetting surface. They established a two-dimensional frequency-domain and time-domain nonlinear hydroelastic mechanics theory to analyze the impact of second-order wave forces on the motion and structural dynamic response of floating structures during navigation or mooring. Chen et al. [6] derived the three-dimensional, second-order hydroelastic mechanics relationship for moored floating structures based on the aforementioned theory, as well as the generalized form of the second-order hydrodynamic coefficients in the three-dimensional nonlinear hydroelastic analysis of sailing speed. The analysis included the linearization method for symmetric anchor chain systems and established three-dimensional linear and nonlinear frequency-domain hydroelastic motion equations for moored floating structures, obtaining nonlinear motion responses and nonlinear structural dynamic responses. Low-frequency resonances persist in elastic mode responses, and coupling can occur, leading to higher peak values. From the contribution of second-order wave forces to the generalized hydrodynamic forces, it is evident that the difference frequency components significantly outweigh the harmonic frequency components. Under high sea conditions, the primary contribution to second-order drift forces is due to the second-order rotation of the rigid body. The methods for handling steady flow fields were summarized by Tian et al. [7], and a technique for eliminating singularities by distributing virtual sources and sinks was proposed. This technique enables the solution of high-order partial differential equations of velocity potential in the non-uniform steady wave flow field of a sailing vessel through numerical methods. Building upon the consideration of the steady wave flow field, they employed the moving pulse source Green function to calculate the linear and nonlinear hydroelastic responses of homogeneous ships and catamarans with small waterlines under non-uniform steady wave flow conditions. By comparing various nonlinear hydrodynamic contributions, it was determined that the nonlinear forces resulting from instantaneous wet surface changes are particularly significant.

Currently, potential flow theory is primarily utilized for calculating and analyzing the hydroelastic responses of various floating structures. However, due to its challenges in numerically simulating phenomena such as wave breaking and slamming, and the neglect of fluid viscosity, it cannot accurately simulate strong nonlinear wave motion and the significant structural movements and deformations it causes, nor can it precisely capture pressure changes near the structure's walls [8]. Compared to other methods, Computational Fluid Dynamics (CFD) coupled with the Finite Element Method (FEM) for fluid–structure interaction yields greater precision and accuracy in depicting changes in velocity and pressure fields within the flow domain. It also excels at capturing nonlinear phenomena at free surfaces, offering robust technical support for addressing nonlinear wave-induced vibrations and other hydroelastic problems [9]. Depending on whether the deformation displacement of the structure feeds back into the flow field, fluid–structure interaction problems can be classified into unidirectional coupling and bidirectional coupling [10].

In the nonlinear response of hydroelasticity, wave-induced vibration and slamming flutter phenomena are typically studied. Fang et al. [11] discovered that computational fluid dynamics methods more accurately reflect the simulated flow field when calculating the hydrodynamic effects of waves on navigating vessels, as opposed to potential flow calculations. Additionally, the results obtained from CFD methods are more precise. Ley et al. [12] developed a unidirectional and bidirectional coupled system based on the Reynolds-averaged Navier–Stokes (RANS) equations for CFD and dynamic FEM. He demonstrated that coupling CFD methods with dynamic FEM can effectively predict loads induced by regular and irregular waves. Seng [13] used OpenFoam to develop a coupling method, employing beam models to calculate hydroelastic responses, and the results showed that the CFD method has good accuracy in estimating the hydroelastic response of structures. Tomoki et al. [14] utilized CFD-FEM technology to analyze the hydroelastic response of a 6600TEU containership and calculated the variations in section moments within the hull. However, the fluid–structure interaction was limited to unidirectional coupling, as bidirectional data exchange between the fluid and structure was not considered. Lakshmynarayanana et al. [15,16] numerically calculated the hydroelasticity of a barge and S-175 containership, respectively, using CFD-FEM methods, finding that this method can fully capture nonlinear factors in the flow field, and the calculated results are closer to experimental values than traditional potential flow elasticity methods. Jiao et al. [17] presented a co-simulation method for the nonlinear hydroelastic response of ships in severe waves with two-way CFD-FEM coupling. S-175 containership tests validated accuracy in modeling motions, loads, and hydroelastic responses. Vijith [18], Lu [19], and Li [20] adopted the same two-way CFD-FEM method to calculate the wave-induced vibration and slamming flutter responses of ultra-large containerships. Hydroelastic tests of large containerships were conducted in a seakeeping tank to measure vertical bending moments, and the test results well validated the preceding numerical results [21,22].

This paper employs a large containership test model, designed and processed by Dr. Si [23] from the China Ship Scientific Research Center, and selects a half-frequency nonlinear wave-induced vibration condition. The study investigates the motion and load response of a large containership under nonlinear wave-induced vibration using both the proprietary three-dimensional hydroelastic software THAFTS V3.0 and the commercial software STARCCM+ 2206 & ABAQUS 6.14 coupled solver, and compares the findings with the model test results.

## 2. Motion and Load Analysis Method for Large Containerships

### 2.1. Nonlinear Hydroelastic Analysis Method in Frequency Domain

Considering the second-order small quantity of the rigid body motion rotation angle of a ship in a wave and the influence of this second-order quantity $\vec{r}\prime = (x\prime, y\prime, z\prime)$ and instantaneous wet surface change on the hydrodynamic load, the total displacement of any point of the hull structure in the coordinate system of the follower can be expressed as:

$$\vec{u} = \sum_{r=1}^{m} \vec{u}_r p_r^{(1)}(\mathrm{t}) + \sum_{r=1}^{m} \vec{u}_r p_r^{(2)}(\mathrm{t}) + \mathbf{H}\vec{r}\prime + \mathbf{R}\sum_{r=7}^{m} \vec{u}_r p_r^{(1)}(\mathrm{t}) \tag{1}$$

Among them, the linear solution $p_r^{(1)}(\mathrm{t})$ is the main coordinate, and the nonlinear response $p_r^{(2)}(\mathrm{t})$ is the main coordinate. The two matrices related to rigid body rotation are the following:

$$\mathbf{R} = \begin{bmatrix} 0 & -\theta_6 & \theta_5 \\ \theta_6 & 0 & -\theta_4 \\ -\theta_5 & \theta_4 & 0 \end{bmatrix}, \mathbf{H} = \frac{1}{2}\begin{bmatrix} -(\theta_5^2+\theta_6^2) & 0 & 0 \\ 2\theta_4\theta_5 & -(\theta_4^2+\theta_6^2) & 0 \\ 2\theta_4\theta_6 & 2\theta_5\theta_6 & -(\theta_4^2+\theta_5^2) \end{bmatrix} \tag{2}$$

According to the three-dimensional hydroelasticity theory, the pressure acting on the instantaneous wet surface of the hull is expanded to two orders through perturbation. Considering the hydrodynamics of the second order, the first-order component of the generalized hydrodynamic force acting on the hull can be expressed as:

$$Z_r(t) = Z_r^{(0)} + Z_r^{(1)}(t) + Z_r^{(2)}(t) \tag{3}$$

In the formula, $Z_r^{(0)}$, $Z_r^{(1)}(t)$, and $Z_r^{(2)}(t)$ are the generalized static fluid force, generalized first-order fluid force, and generalized second-order fluid force, respectively.

The generalized static force of fluid $Z_r^{(0)}$ is the sum of the generalized steady buoyancy and the generalized steady pressure of the flow field, that is

$$Z_r^{(0)} = \rho \iint_{\overline{S}} \vec{n} \cdot \vec{u}_r \left[ gz\prime + \frac{1}{2}(W^2 - U^2) \right] dS \tag{4}$$

The first-order flow $Z_r^{(1)}(t)$ in the broad sense can be expressed as:

$$Z_r^{(1)}(t) = F_r^{(1)}(t) + D_r^{(1)}(t) + R_r^{(1)}(t) + \Delta R_r(t) \tag{5}$$

$$F_r^{(1)}(t) = \rho \iint_{\overline{S}} \vec{n} \cdot \vec{u}_r \left( \frac{\partial}{\partial t} + \vec{W} \cdot \nabla \right) [\phi_I(t) + \phi_D(t)] dS \tag{6}$$

$$D_r^{(1)}(t) = \sum_{k=1}^{m} \rho \iint_{\overline{S}} \vec{n} \cdot \vec{u}_r \left( \frac{\partial}{\partial t} + \vec{W} \cdot \nabla \right) \phi_k(t) dS \tag{7}$$

$$R_r^{(1)}(t) = \rho \iint_{\overline{S}} \vec{n} \cdot \vec{u}_r \left[ gw + \frac{1}{2}(\vec{u}_r \cdot \nabla) W^2 \right] dS \tag{8}$$

$$\Delta R_r(t) = \rho \iint_{\overline{S}} (\mathbf{R}\vec{n}) \cdot \vec{u}_r \left[ gz\prime + \frac{1}{2}(W^2 - U^2) \right] dS \tag{9}$$

where $F_r^{(1)}(t)$, $D_r^{(1)}(t)$, and $R_r^{(1)}(t)$, respectively, represent the generalized first-order wave excitation force, the generalized first-order radiation force, and the generalized first-order restore force, and $\Delta R_r(t)$ is the first-order fluid restore force caused by the large angular rotation of the rigid body.

The generalized second-order fluid mechanics can be expressed as:

$$Z_r^{(2)}(t) = F_r^{(2)}(t) + D_r^{(2)}(t) + E_r^{(2)}(t) + S_r^{(2)}(t) + \Delta Z_r^{(2)}(t) \tag{10}$$

$$\begin{aligned} F_r^{(2)}(t) &= \rho\iint_{\overline{S}} \left[\left(\mathbf{R}\vec{n}\right)\cdot\vec{u}_r + (\vec{n}\cdot\vec{u}_r)(\vec{u}_r\cdot\nabla)\right]\left(\frac{\partial}{\partial t}+\vec{W}\cdot\nabla\right)[\phi_I(t)+\phi_D(t)]dS \\ &+\rho\iint_{\overline{S}} \vec{n}\cdot\vec{u}_r\frac{1}{2}[\nabla\phi_I(t)+\nabla\phi_D(t)]^2 dS \end{aligned} \tag{11}$$

$$\begin{aligned} D_r^{(2)}(t) &= \sum_{k=1}^{m}\rho\iint_{\overline{S}} \left[\left(\mathbf{R}\vec{n}\right)\cdot\vec{u}_r + (\vec{n}\cdot\vec{u}_r)(\vec{u}_r\cdot\nabla)\right]\left(\frac{\partial}{\partial t}+\vec{W}\cdot\nabla\right)\phi_k(t)dS \\ &+\sum_{k=1}^{m}\sum_{l=1}^{m}\frac{1}{2}\rho\iint_{\overline{S}}(\vec{n}\cdot\vec{u}_r)\nabla\phi_k(t)\cdot\nabla\phi_l(t)dS \end{aligned} \tag{12}$$

$$E_r^{(2)}(t) = \sum_{k=1}^{m}\rho\iint_{\overline{S}}(\vec{n}\cdot\vec{u}_r)\nabla[\phi_I(t)+\phi_D(t)]\cdot\nabla\phi_k(t)dS \tag{13}$$

$$\begin{aligned} S_r^{(2)}(t) &= \rho\iint_{\overline{S}}(\mathbf{R}\vec{n})\cdot\vec{u}_r\left[gw+\frac{1}{2}(\vec{u}_r\cdot\nabla)W^2\right]dS \\ &+\rho\iint_{\overline{S}}(\mathbf{H}\vec{n})\cdot\vec{u}_r\left[gz\prime+\frac{1}{2}(W^2-U^2)\right]dS \\ &+\rho\iint_{\overline{S}}(\vec{n}\cdot\vec{u}_r)(\mathbf{H}\vec{r}\prime+\mathbf{R}\vec{u}_d)\cdot\nabla(gz\prime+\frac{1}{2}W^2)dS \end{aligned} \tag{14}$$

$$\Delta Z_r^{(2)}(t) = \rho\iint_{\Delta S}\vec{n}\cdot\vec{u}_r\left[(\frac{\partial}{\partial t}+\vec{W}\cdot\nabla)\phi+\frac{1}{2}(W^2-U^2)+\frac{1}{2}(\vec{u}_r\cdot\nabla)W^2+g(z\prime+w))\right]dS \tag{15}$$

where $F_r^{(2)}(t)$ and $D_r^{(2)}(t)$ are the generalized second-order wave excitation force and generalized second-order radiation force, respectively. $E_r^{(2)}(t)$ is the coupling term of radiation potential, incident potential, and reflected potential in the second-order force; $S_r^{(2)}(t)$ is the contribution force of static pressure of the first-order flow field to the generalized second-order fluid force; $\Delta Z_r^{(2)}(t)$ is the second-order force caused by the instantaneous wet surface change; and $\Delta S = S - \overline{S}$ is the difference between the instantaneous wet surface and the average wet surface.

### 2.2. CFD-FEM Coupling Method

The finite volume method (Finite Volume Method, VOF) is a widely used discretization technique in CFD. In STARCCM+, the "separated flow" solver is used to solve the velocity and pressure fields separately. A two-phase flow model is employed for simulation, without considering the compressibility of air and water. The governing equations for the flow field are the time-averaged Navier–Stokes equations (RANS equations) and the continuity equation:

$$\begin{cases} \rho\left(\frac{\partial\overline{u}_i}{\partial t}+\frac{\partial\left(\overline{u_i u_j}\right)}{\partial x_j}\right) = -\frac{\partial\overline{p}}{\partial x_i}+\mu\nabla^2\overline{u}_i+\frac{\partial\left(-\rho\overline{u_i'u_j'}\right)}{\partial x_j} \\ \frac{\partial\overline{u}_j}{\partial x_j}=0 \end{cases} \tag{16}$$

where $u$ is velocity, $\rho$ is fluid density, $p$ is pressure, $\mu$ is the viscosity coefficient, $\overline{u}$ is the time average of velocity $u$, and $\overline{p}$ is the time average of pressure $p$. $-\rho\overline{u_i'u_j'}$ is the Reynolds stress, a second-order tensor. The k-ε model is used to establish a supplementary relationship between turbulent stress and mean velocity, making the control equations closed. The free surface is captured using the VOF method. The computational domain of the flow field is divided using a cut-cell mesh, and an overlapping mesh is applied to the region containing the hull, with the basin moving along with the hull.

The wave forcing method in STARCCM+ is used to eliminate the wave. Compared with the traditional damping wave elimination method, this method significantly reduces the size of the calculation domain and improves the calculation efficiency.

FEM uses dynamic motion equations to solve the motion response of the structure:

$$M\ddot{u} + C\dot{u} + Ku = F \quad (17)$$

where $M$ is the mass matrix of the structure, $C$ is the damping matrix of the structure, $K$ is the stiffness matrix of the structure, $u$ is the displacement of the structure, $\dot{u}$ is the velocity of the structure, $\ddot{u}$ is the acceleration of the structure, and $F$ is the external load of the structure.

## 3. Comparison Between Model Test Results and Numerical Results

### 3.1. Model Test Setup

The large containership is converted into a test model according to the scale ratio λ = 1/77. The basic data are shown in Table 1. The left column shows the full-scale ship data, and the right column shows the scaled model data. The test model is shown in Figure 1.

**Table 1.** Basic data of a large containership.

| Parameter | Ship | Model (λ = 1/77) |
|---|---|---|
| Length between perpendiculars $L_{BP}$ (m) | 383.0 | 4.974 |
| Breadth B (m) | 58.6 | 0.761 |
| Depth H (m) | 30.5 | 0.396 |
| Aft draft $T_a$ (m) | 16.7 | 0.217 |
| Fore draft $T_f$ (m) | 15.2 | 0.197 |
| Displacement Δ (ton) | 260,602.9 | 0.557 |
| Vertical center of gravity $Z_g$ (m) | 27.574 | 0.358 |
| Longitudinal center of gravity $X_g$ (m) | 182.714 | 2.373 |

**Figure 1.** Test model of the large containership.

For the 1:77 scale ship model, a new type of U-shape backbone, seen in Figure 2, was designed to model the bending of the containership in waves. According to the similitude principle, a kind of ABS PC plastic plate was chosen to construct the U-shape backbone. The Young's modulus of ABS PC is 2.4 GPa, which is about 1/85 of the steel. The Poisson ratio is 0.3897. The backbone beam in blue is connected to the ship hull through the transverse wood of a rectangular cross-section.

### 3.2. Numerical Setup

In this work, numerical simulations are conducted using the potential flow model and CFD model based on the unsteady Reynolds-averaged Navier–Stokes (RANS) equations solver. Figures 2–4 illustrate the finite element structure model of the large containership, the potential flow model, and the CFD computational domain. The structure employed for the CFD-FEM coupling model satisfies the quality and stiffness distribution criteria of the hull structure, as presented in Figure 5. To accurately simulate the vibration characteristics of a hull girder, the requirements of stiffness similitude on vertical bending were met completely. In the 2D beam model, the corresponding stiffness was adjusted by changing the size of the beam cross-section.

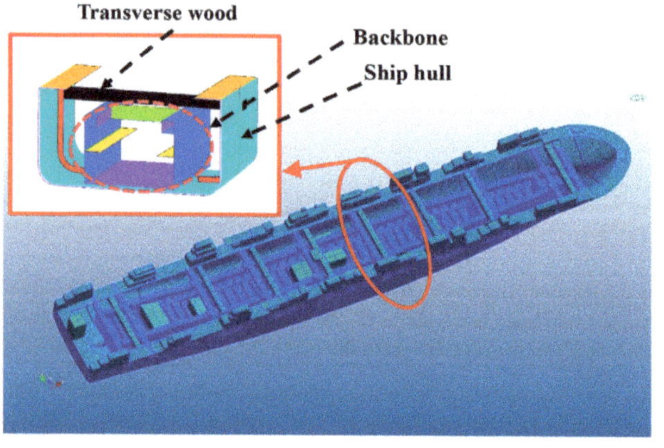


**Figure 2.** Finite element model of test model for large containership in THAFTS.

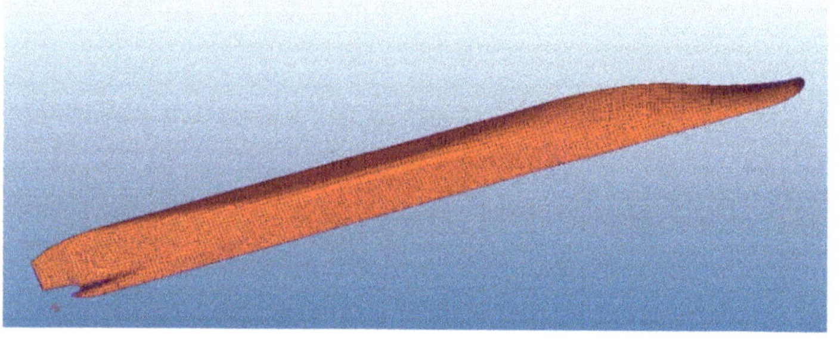

**Figure 3.** Hydrodynamic panels for large containerships in THAFTS.

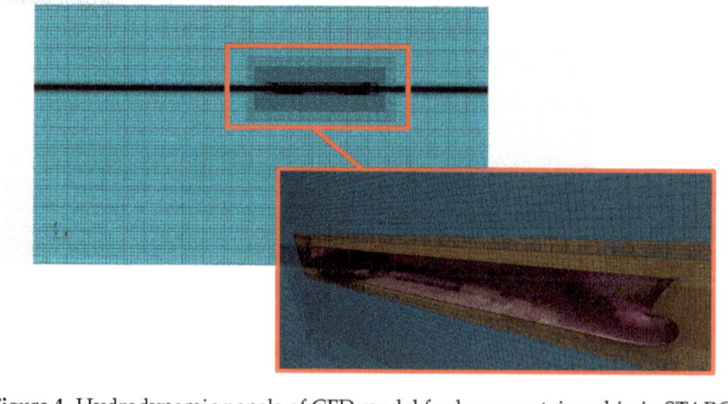

**Figure 4.** Hydrodynamic panels of CFD model for large containership in STARCCM+.

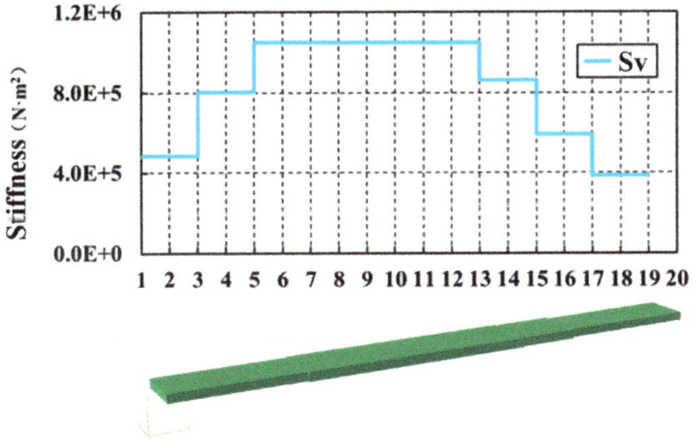


**Figure 5.** Structural grids of CFD model for large containership in STARCCM+.

To examine grid sensitivity, three systematically refined meshes for both the potential flow model and CFD model are employed, with a uniform refinement ratio of hi + 1/hi = 1.414, where hi and hi + 1 represent the grid spacings of two successive levels. The grid parameters for the two numerical models are summarized in Table 2.

**Table 2.** The mesh independence study.

| No. | hi/h1 (i = 1,2,3) | FD Mesh (Hull) | CFD Mesh (Hull) | CFD Mesh (Computational Domain) |
|---|---|---|---|---|
| 1 | 1 | $2.6 \times 10^3$ | $1.0 \times 10^4$ | $2.05 \times 10^5$ |
| 2 | 1.414 | $3.7 \times 10^3$ | $1.4 \times 10^4$ | $2.9 \times 10^6$ |
| 3 | 2 | $5.3 \times 10^3$ | $2.0 \times 10^4$ | $4.1 \times 10^6$ |

The time history curve of wave elevation with potential flow and viscous flow numerical results are depicted in Figure 6. The potential flow frequency domain calculation results in strong agreement among the three grids. In the frequency domain analysis model, the wavelength corresponding to the resonance frequency comprises a grid of at least five lengths. Grid 2 meets the calculation accuracy requirements based on the minimum number of meshes included in the minimum wavelength. As for the CFD simulation, good agreement is found among the results of Grid 2 and Grid 3, while the wave amplitude in Grid 1 is slightly reduced. The convergence ratio decreases with an increase in the mesh number, which means the numerical model is monotonically converging. According to the analysis of wave elevation, Grid 2 is selected as an appropriate resolution since the wave profiles are captured better than Grid 1, while the calculation is less demanding than Grid 3 for the CFD simulations.

### 3.3. Nonlinear Case

The modal shapes corresponding to this resonant frequency reveal significant vertical bending deflections at the two-node mode, indicating potential areas of concern for structural integrity. Additionally, the CFD model, as shown in Figures 4 and 5, captures the complex fluid–structure interaction, enabling a more accurate prediction of the ship's response to wave loading. This detailed modeling approach is crucial for understanding and mitigating the effects of wave-induced springing on large containerships. The resonant frequency of the vertical bending springing at two nodes of a containership model traveling in the wave basin at a speed of 23 knots (1.348 m/s at model scale) is 3.589 Hz (22.55 rad/s). The incident wave frequency that triggers this resonance is generally 9.68 rad/s, with a wavelength of 0.658 m. However, due to the constraints on maximum amplitude when generating short waves in the seakeeping tank, the incident wave frequency corresponding to half the resonant frequency of the vertical bending wave-induced vibration at two nodes, which is 6.06 rad/s (wavelength equals 1.677 m), is chosen as the wave condition for the bow wave test in the tank. For the full-speed test conditions of the containership model, refer to Table 3.

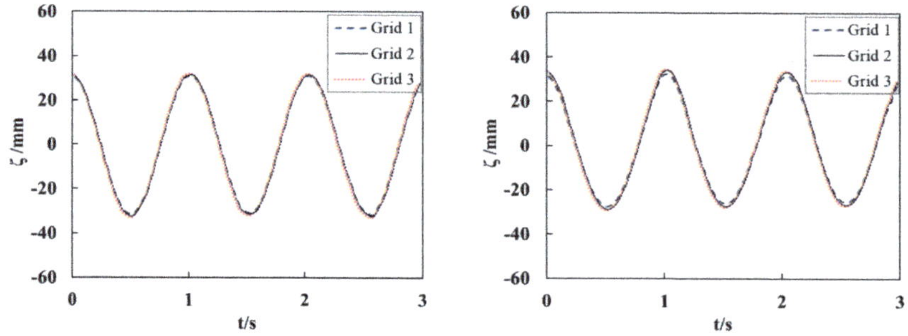


(**a**) Potential flow model (**b**) CFD simulations

**Figure 6.** Effect of the number of grid cells.

**Table 3.** The case of ship model test for nonlinear wave-induced springing.

| Case | Speed | | Wave height | | Wavelength | | Direction (°) |
|---|---|---|---|---|---|---|---|
| | Ship (kn) | Model (m/s) | Ship (m) | Model (mm) | Ship (m) | Model (m) | |
| No. 1 | 23.0 | 1.348 | 5.0 | 64.9 | 129.1 | 1.677 | 180 |

Based on linear hydroelasticity theory, and by comparing frequency domain calculations with CFD results, the table below (Table 4) illustrates the comparison between the linear springing frequency of the containership model and the wave frequency under test conditions.

**Table 4.** The linear springing frequency and the test condition frequency (Hz).

| Item | Calculate the Frequency of Linear Springing | Wave Frequency in Test | Encounters Frequency of Model Test $\omega_e$ | 2 $\omega_e$ | Error |
|---|---|---|---|---|---|
| Values | 3.589 | 0.98 | 1.80 | 3.60 | −0.31% |

*3.4. Discussion*

The three-dimensional hydroelastic mechanics frequency-domain nonlinear analysis method and the CFD-FEM coupled hydroelastic method were applied to calculate the conventional vertical motion and vertical bending moment in nonlinear wave-induced vibration.

The wave height, time history of containership motion, spectrum, data analysis, and comparison with potential flow and viscous flow numerical results are depicted in Figures 7 and 8. Here, "TEST" represents the test results, "FD" represents the potential flow frequency domain calculation results, and "CFD" represents the viscous flow calculation results with STARCCM+.

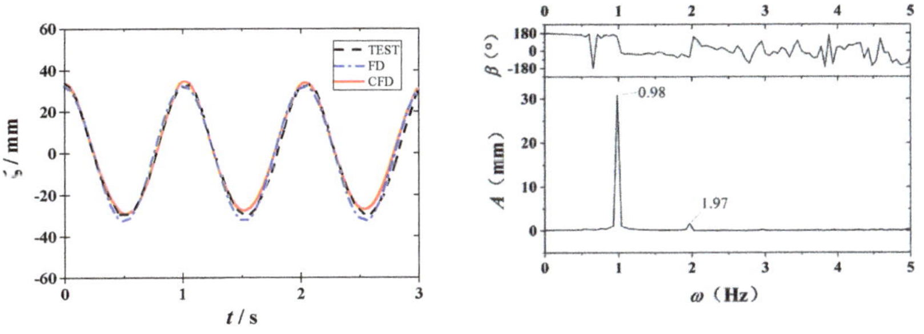


(**a**) Time history curve of wave elevation (**b**) FFT spectrum of wave elevation

**Figure 7.** Wave test data analysis.

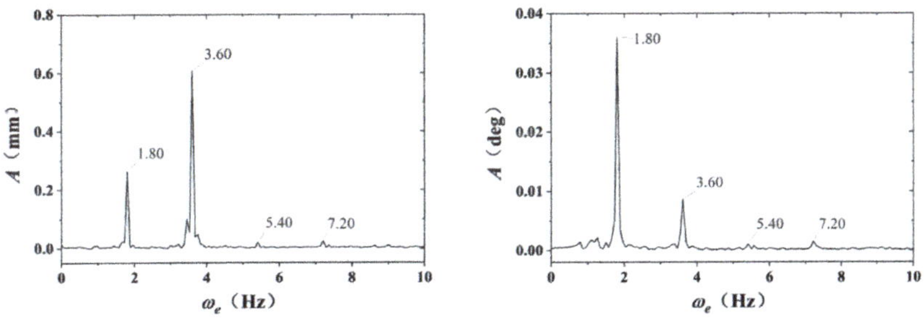


(**a**) FFT spectrum diagram at heave motion (**b**) FFT spectrum diagram at pitch motion

**Figure 8.** Data analysis of motion test of large containership test model.

The incident wave serves as the input condition for the entire ship model measurement system, with a frequency of 0.98 Hz (wavelength 1.677 m). Nevertheless, the test wave's high-frequency measurement information reveals a frequency of 1.97 Hz, which is twice that of the incident wave. This doubling is attributed to the unavoidable higher-order harmonic phenomenon inherent in the wave-making process within the water tank. However, these higher-order harmonics are quite minor. For wave-making, the numerical results align well with experimental outcomes, ensuring the precision of subsequent calculations. Given that nonlinear wave-induced vibrations arise from the resonant frequencies of containerships in the water, the spectrum diagram from heave and pitch measurements displays multiple vibration peaks at frequencies such as 1.80 Hz, 3.60 Hz (2×), 5.40 Hz (3×), and 7.20 Hz (4×). Among these, the encounter frequency for the containership at this speed is 1.80 Hz. Since the encounter frequency is proximate to half of the linear wave-induced vibration frequency, it is susceptible to inducing a 2× harmonic nonlinear wave-induced vibration, as per second-order nonlinear effects.

The nonlinear frequency domain hydroelastic analysis method is employed to calculate the motion and structural dynamic response of the ship model under the influence of regular waves with a frequency corresponding to the test encounter frequency within the Kelvin wave field. Additionally, the motion and section bending moment within the structure under these conditions are determined using CFD-FEM coupling calculations, as depicted in Figures 9–11. Both methods demonstrated high consistency in motion and load under head wave conditions. When combined with data comparison, it is confirmed that both methods are effective for evaluating the nonlinear wave-induced vibration response of regular waves.

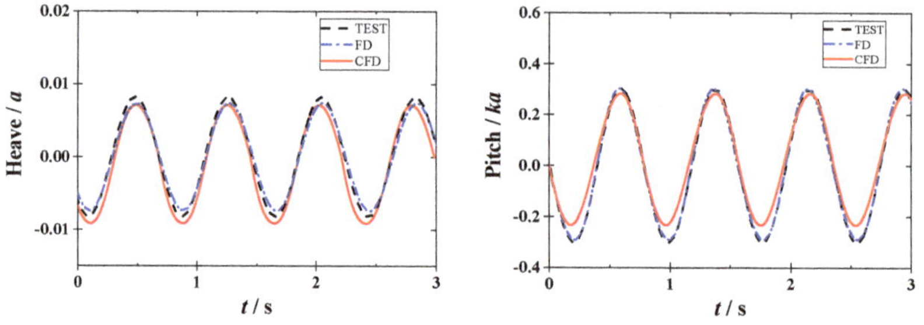


**Figure 9.** Comparison of numerical results and model test for the motions of large containership at wave frequency.

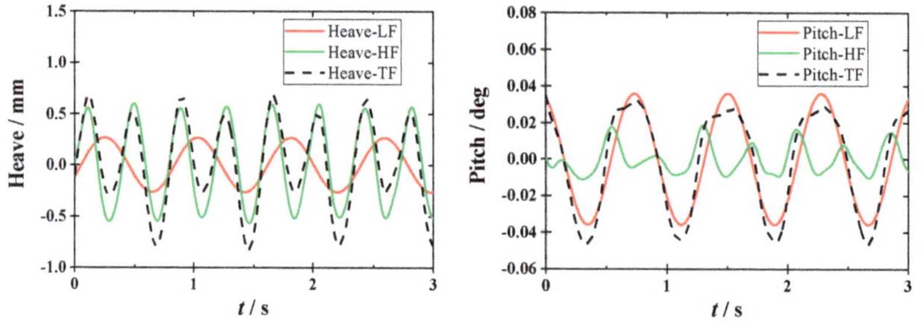


**Figure 10.** Comparison of linear and nonlinear numerical simulations for the motions of large containership.

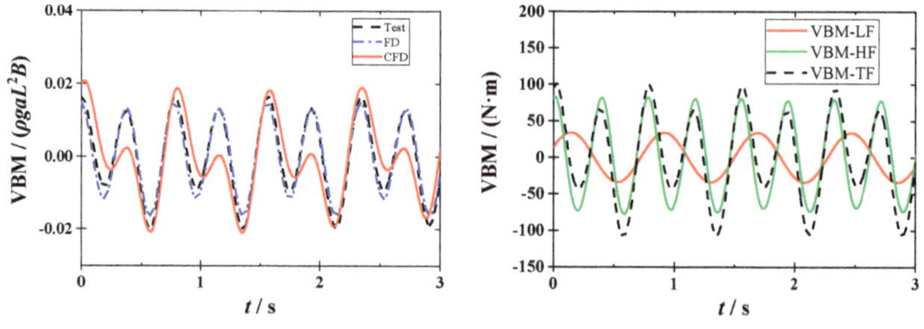


**Figure 11.** Comparison of numerical results and model test of VBM of large containership.

From the comparison of the numerical results of the large containership test model's motions at the wave frequency shown in Figure 8 with the experimental results, it can be seen that both the frequency domain calculation method for potential flow theory and the CFD viscous flow calculation method yield numerical simulations of rigid body motion that are relatively close to the experimental results. Since the given wave incidence frequency components are fixed, the time course periods of all three methods can be effectively guaranteed. However, there are some subtle differences in amplitude, mainly due to slight variations in the damping settings of the numerical system compared to those measured in the model tests, as well as the more pronounced effect of the viscous damping term under the CFD-FEM coupling method.

Nonlinear wave-induced vibrations appear as high-frequency bending responses in the ship's structure, resulting in higher-order harmonic components, such as those at frequencies 2× and 3×, in the motions and section vertical bending moments during operation, as shown in Figures 10 and 11.

Figure 9 depicts the heave and pitch motions of the large containership over time, showcasing a comparison of results from different numerical approaches. The horizontal axis represents time t in seconds, spanning from 0 to 3 s, while the vertical axis indicates heave displacement in millimeters and pitch angle in degrees. Three distinct curves are presented as follows: -LF (red solid line) depicts low-frequency linear response, -HF (green solid line) depicts high-frequency nonlinear response, and -TF (green long-dashed line) depicts total-frequency response, respectively. This mutually validates with Figure 8, including the amplitude and frequency.

All curves exhibit periodic oscillations, which is consistent with the typical heave and pitch motion characteristics of a ship under wave excitation. However, slight differences in amplitude and phase among the curves reflect the disparities between the frequencies. Linear-based methods (e.g., Heave-LF, Pitch-LF) can only calculate and describe the responses under a single regular wave. Since the encounter frequency (1.80 Hz) under this wave is exactly half of the first-order resonant frequency (3.60 Hz) of the ship model, through the superposition of sum frequencies, the nonlinear calculated high frequency response (e.g., Heave-HF, Pitch-HF) is exactly the response at the resonant frequency. Such comparisons are crucial for validating numerical algorithms and understanding the ship's nonlinear springing behavior in waves, aiding in the design and optimization of containerships for better hydrodynamic performance.

As the frequency of wave encounters and its multiples coincide with the resonant frequencies of the ship's structure after accounting for fluid-added mass, nonlinear wave-induced vibrations are triggered by waves. Considering the section vertical bending moment (VBM) as an example, as illustrated in Figure 11, the outcomes of the nonlinear frequency-domain hydroelastic analysis, which employs potential flow theory, are quite close to the experimental data. Nevertheless, due to phase differences in the low-frequency and high-frequency responses, the coupled CFD-FEM calculation method exhibits some discrepancies in the combined response.

## 4. Conclusions

This study systematically employs nonlinear frequency-domain hydroelastic theory and the CFD-FEM coupled hydroelastic calculation method to address the nonlinear, multi-frequency wave-induced vibration response of containerships in regular waves, with comprehensive validation against self-propelled model tests.

Specifically, the research focuses on simulating the design-speed head wave scenario in a wave basin, where motion and load response curves are acquired through experimental measurements. By implementing spectral analysis and time-history band-pass filtering, the response components are decomposed into low-frequency (wave frequency) and high-frequency (wave frequency multiples, i.e., springing resonance frequency) parts, enabling a detailed discussion of vibration mechanisms.

Based on potential flow theory, the three-dimensional frequency-domain nonlinear hydroelastic analysis method considers the second-order hydrodynamic effects induced by large-angle rigid-body rotations and instantaneous wet surface variations. In contrast, the CFD method can effectively account for the nonlinear factors of the flow field. By coupling with the FEM to solve the ship's hydroelastic problem, not only can the effects of instantaneous wet surface and large angle rotation on the second-order wave forces in potential flow theory be captured, but also the effects of structural deformation on the flow field can be further carefully considered. Thus, the CFD-FEM coupling approach integrates the finite volume method (VOF) for free-surface capture and dynamic FEM for structural response, achieving two-way fluid–structure interaction.

Key findings reveal that both methods demonstrate high consistency in predicting vertical heave/pitch motions and vertical bending moments (VBMs) under head waves. Notably, these two numerical methods show superior accuracy in capturing nonlinear phenomena, such as high-frequency harmonic components (two times and three times of wave frequencies) in VBM responses, which arise from the resonance between encounter frequencies and structural natural frequencies (e.g., 1.80 Hz encounter frequency triggering 3.60 Hz springing resonance).

Experimental validation confirms that the frequency-domain method based on potential flow theory efficiently predicts rigid-body motions, while CFD-FEM coupling better accounts for viscous damping effects and structural-deformation-induced flow field changes. The discrepancy in combined responses (low-frequency + high-frequency) between methods is attributed to phase differences in dynamic responses, highlighting the necessity of integrating both approaches for comprehensive fatigue assessment. These results establish the effectiveness of nonlinear hydroelastic analysis and CFD-FEM coupling in evaluating wave-induced springing, providing critical technical support for hydrodynamic design optimization and fatigue strength evaluation of ultra-large containerships in severe sea conditions. Additionally, the frequency-domain nonlinear method has a fast calculation speed, making it more suitable for the preliminary design of large ships.

**Author Contributions:** Validation, L.L.; Investigation, Y.L.; Writing—review & editing, Q.J. All authors have read and agreed to the published version of the manuscript.

**Funding:** This paper is supported by the Basic Research of Jiangsu (No. BK20243019), the CSSRC-20,000TEU JIP, and the Hubei Provincial Natural Science Foundation of China (No. 2023AFB126).

**Data Availability Statement:** The original contributions presented in this study are included in the article. Further inquiries can be directed to the corresponding author.

**Conflicts of Interest:** The authors declare no conflict of interest.

## References

1. Hirdaris, S.E.; Bai, W.; Dessi, D.; Ergin, A.; Gu, X.; Hermundstad, O.A.; Huijsmans, R.; Iijima, K.; Nielsen, U.D.; Parunov, J.; et al. Loads for use in the design of ships and offshore structures. *Ocean Eng.* **2014**, *78*, 131–174. [CrossRef]
2. Hirdaris, S.E.; Lee, Y.; Mortola, G.; Incecik, A.; Turan, O.; Hong, S.Y.; Kim, B.W.; Kim, K.H.; Bennett, S.; Miao, S.H.; et al. The influence of nonlinearities on the symmetric hydrodynamic response of a 10,000 TEU container ship. *Ocean Eng.* **2016**, *111*, 166–178. [CrossRef]
3. Wu, Y. Hydroelasticity of Floating Bodies. Ph.D. Thesis, Brunel University, Uxbridge, UK, 1984.
4. Price, W.G.; Wu, Y. Hydroelasticity of marine structures. In *Theoretical and Applied Mechanics*; Elsevier: Amsterdam, The Netherlands, 1985; pp. 311–337.
5. Wu, Y.; Maeda, H.; Kinoshita, T. The Second Order Hydrodynamic actions on a flexible body. *J. Inst. Ind. Sci. Univ. Tokyo* **1997**, *49*, 8–19.
6. Chen, X.; Wu, Y.; Cui, W.; Tang, X. Nonlinear hydroelastic analysis of a moored floating body. *Ocean Eng.* **2003**, *30*, 965–1003. [CrossRef]
7. Tian, C.; Wu, Y. The second-order hydroelastic analysis of a swath ship moving in large-amplitude waves. *J. Hydrodyn. Ser. B* **2006**, *18*, 631–639. [CrossRef]
8. Liu, G.; Sun, Z.; Li, H.; Zou, L.; Zhen, H. A coupled CFD-FEM method for computation of ship hydroelasticity. *Shipbuild. China* **2022**, *63*, 89–101.
9. Zhang, G.; Wang, S.; Sun, Z.; Xiao, Q. Research developments in numerical methods of fluid-structure interactions in naval architecture and ocean engineering. *Chin. J. Ship Res.* **2022**, *17*, 52–73.

10. Liu, Y.; Zhu, R.; Cheng, Y. Numerical simulation of hydroelastic responses of floating structure based on CFD-FEM method. *Ocean Eng.* **2020**, *38*, 24–32.
11. Fang, Z.; Zhu, R.; Miao, G. Numerical simulation on radiation problems of moving vessels in numerical wave tank. *Chin. J. Hydrodyn.* **2011**, *27*, 515–524.
12. Ley, J.; Moctar, O. An enhanced one-way coupling method to predict elastic global hull girder loads. In Proceedings of the 33rd International Conference on Ocean, Offshore and Arctic Engineering OMAE2014-24199, San Francisco, CA, USA, 8–13 June 2014.
13. Seng, S. *Slamming and Whipping Analysis of Ships*; Technical University of Denmark: Copenhagen, Denmark, 2012.
14. Tomoki, T.; Sadaoki, M.; Masayoshi, O. A numerical simulation method for predicting global and local hydroelastic response of a ship based on CFD and FEA coupling. *Mar. Struct.* **2018**, *59*, 368–386.
15. Lakshmynarayanana, P.; Temarel, P. Application of CFD and FEA coupling to predict dynamic behavior of a flexible barge in regular head waves. *Mar. Struct.* **2019**, *65*, 308–325. [CrossRef]
16. Lakshmynarayanana, P.; Hirdaris, S. Comparison of nonlinear one- and two-way FFSI methods for the prediction of the symmetric response of a containership in waves. *Ocean Eng.* **2020**, *203*, 107179. [CrossRef]
17. Jiao, J.; Huang, S.; Wang, S.; Soares, C.G. A CFD-FEA two-way coupling method for predicting ship wave loads and hydroelastic responses. *Appl. Ocean Res.* **2021**, *117*, 102919. [CrossRef]
18. Vijith, P.; Rajendran, S. Hydroelastic effects on the vertical bending moment of a container ship in head and oblique seas. *Ocean Eng.* **2023**, *285*, 115385.
19. Lu, L.; Ren, H.; Li, H.; Zou, J.; Chen, S.; Liu, R. Numerical method for whipping response of ultra large container ships under asymmetric slamming in regular waves. *Ocean Eng.* **2023**, *287*, 115830. [CrossRef]
20. Li, P.; Xie, H.; Liu, F.; Yu, P.; Liu, X. Numerical approach for predicting the slamming loads in oblique long waves considering the wave-body interface. *Ocean Eng.* **2022**, *262*, 112089. [CrossRef]
21. Kim, S.; Bouscasse, B.; Ducrozet, G.; Delacroix, S.; De Hauteclocque, G.; Ferrant, P. Experimental investigation on wave-induced bending moments of a 6, 750-TEU containership in oblique waves. *Ocean Eng.* **2023**, *284*, 115161. [CrossRef]
22. Jiao, J.; Ren, H.; Sun, S.; Adenya, C.A. Experimental investigation of wave-induced ship hydroelastic vibrations by large-scale model measurement in coastal waves. *Shock. Vib.* **2016**, *2016*, 1–14. [CrossRef]
23. Si, H. Research on Springing and Whipping Slamming of Ultra-Large Container Ships. Ph.D. Thesis, China Ship Scientific Research Center, Wuxi, China, 2023.

Journal of
*Marine Science and Engineering*

MDPI

*Article*

# Effect of Damping Plate Parameters on Liquid Sloshing in Cylindrical Tanks of Offshore Launch Platforms

**Yuxin Pan [1], Yuanyuan Wang [2], Fengyuan Liu [3] and Gang Xu [1,4,*]**

1 Marine Equipment and Technology Institute, Jiangsu University of Science and Technology, Zhenjiang 212008, China; pyx18360735895@yeah.net
2 Hudong-Zhonghua Shipbuilding (Group) Co., Ltd., Shanghai 201913, China; 18652742046@163.com
3 Makarov College of Marine Engineering, Jiangsu Ocean University, Lianyungang 222005, China; 19851171321@163.com
4 School of Naval Architecture and Ocean Engineering, Jiangsu University of Science and Technology, Zhenjiang 212003, China
* Correspondence: g.xu@just.edu.cn

**Abstract:** To meet the growing demand for space launches and overcome the limitations of land-based launches, the scientific research community is committed to developing safer and more flexible offshore rocket launch technologies. Their core carriers—marine platforms—are directly exposed to the dynamic and variable marine environment. The complex coupling effects of wind, waves, and currents impose severe challenges upon these platforms, causing complex phenomena such as severe rocking. These phenomena pose severe threats to and significantly interfere with the stability and normal execution of offshore rocket launch operations. This study employs CFD simulation software to analyze liquid sloshing within a cylindrical tank, both with and without baffles. Following validation of the natural frequency, the analysis focuses on the suppression effect of different baffle positions and configurations on tank sloshing. The numerical simulation results indicate the following: Incorporating baffles alters the natural frequency of liquid sloshing within the tank and effectively suppresses the free surface motion. The suppression of the wave surface motion improves as the baffle is positioned closer to the free surface and as the number of perforations in the baffle increases. However, when the number of perforations exceeds a certain threshold, further increasing it yields negligible improvement in the suppression of the sloshing wave surface motion.

**Keywords:** tank sloshing; CFD simulation; damping plate

## 1. Introduction

It is widely recognized that marine environmental conditions evolve rapidly, imposing stringent demands on offshore platforms as critical infrastructure. Particularly under harsh metocean conditions, platforms experience complex dynamic loads from wind, waves, and currents—both during stationary positioning and operational phases. Consequently, comprehensive investigation of seaborne satellite launch platforms operating in such challenging ocean environments is imperative.

In recent years, many scholars at home and abroad have studied the fluid characteristics of the phenomenon of liquid shaking in a liquid tank. The phenomenon of sloshing is very relevant for vessels that transport liquid petroleum products—crude/chemical oil tankers—as well as for vessels that transport liquefied petroleum or natural gas—LPG/LNG tankers. Liquid sloshing in the tank not only weakens the stability of the ship and various liquid storage tanks, but also threatens the ship's structural integrity and the safety of its personnel due to the shock loads it induces [1–4]. Because of this, scholars at home and abroad have carried out systematic research on the hydrodynamic characteristics of liquid tank sloshing.

McCarty and Stephens [5] performed numerical simulations of the sloshing of a spherical tank partially filled with liquid, and showed that the higher the free liquid level inside the sphere, the higher the natural frequency of liquid sloshing; and the larger the radius of the tank, the lower the natural frequency of liquid. Francecutto et al. [6] conducted a numerical calculation study on a liquid tank with damping plates in 1996 based on the RANS equation, and the results showed that a liquid tank with damping plates also had a good anti-roll effect. Ahmed F. Abdel Gawad et al. [7] used numerical calculations to change the parameters of a U-shaped liquid tank, and discussed its anti-roll effect in detail, and concluded that the liquid tank should be equipped with a damping structure to control the fluid movement, and a well-designed and tuned liquid tank can be very effective in reducing roll. Abbas Maleki et al. [8] conducted a liquid sloshing study on cylindrical liquid tanks with baffles in 2008 and showed that annular baffles were more effective in reducing sloshing oscillations. Jedediah Morse Store et al. [9] performed an experimental, numerical analysis of water and liquid nitrogen in the liquid tank of a spherical storage tank. Jun Liu et al. [10] analyzed the sloshing performance of various barrel-shaped cylindrical liquid tanks based on the isometric boundary element method, and the results showed that arranging three circular baffles along the excitation direction was the best arrangement to suppress the sloshing of liquids.

Owing to its inherent advantages in handling free surfaces, the meshless particle method has been widely adopted for simulating violent free-surface flows [11–14]. Leng Fei [15] studied the effect of a damping plate on liquid sloshing in a liquid tank based on the SPH method, and the results showed that the presence of a damping plate could make the liquid sloshing return to calm faster. Liu Fu [16] carried out a numerical simulation of liquid sloshing with or without damping plates in a prismatic liquid tank based on the SPH method, and the results showed that the addition of damping plates could reduce the pressure at the top of the liquid tank and make the liquid sloshing amplitude lower. Wu Jianlin et al. [17] discussed the influence of a U-shaped liquid tank structure's damping on the roll reduction performance based on the CFD method, and proposed a method based on numerical calculations to evaluate the roll reduction effect of the tank in regular waves. In the same year, Luo Hanbing [18], based on the open-source software OpenFOAM and a calculation package based on the two-phase flow solver InterFOAM, numerically predicted the inherent period and damping of liquid sloshing in a liquid tank model, and the results showed that a damping plate in the liquid tank would increase the period and damping, and the liquid-level height in the liquid tank would increase significantly.

Mi-An Xue et al. [19] studied viscous liquid sloshing in liquid tanks with internal baffles of different shapes and arrangements based on the three-dimensional (3-D) numerical model NEWTANK, and the results showed that the height of the free liquid level near the liquid bulkhead decreased due to the presence of an annular baffle, and when the annular baffle was close to the free liquid level and the width increased, the annular baffle was more effective in reducing violent liquid sloshing and the presence of the annular baffle caused the peak response frequency to shift to the lower side. Dasgupta A (2011) [20] analyzed the effects of the tank cross-section and longitudinal baffle on transient liquid sloshing of partially filled road tankers based on CFD methods, and the results showed that the addition of longitudinal baffles can significantly reduce the motion of the roll plane. Budiansky (2012) [21] performed numerical simulations of tank sloshing in empty, semi-liquid-filled, and fully liquid-filled spherical tanks, and showed that the natural frequency of liquid sloshing mainly depends on the tank radius and a dimensionless parameter determined in the project. Nema P (2014) [22], based on the finite volume method (FVM)-based fluid volume method (VOF), studied the shaking behavior of a three-dimensional rectangular liquid tank with and without a baffle under external force excitation, showing that the sloshing would become violent under the excitation of the natural frequency under resonance conditions, showing over-turning and causing a serious impact on the roof wall of the tank. A baffle was set in the tank, which acted as a damper and minimized the number of sloshing waves. At the same time, the higher the level of the tank, the greater the complexity of the liquid shaking. Rohit Suyal (2016) [23] conducted a CFD analysis of fuel sloshing inside a cylindrical tank with and without damping baffles at linear acceleration, and the study showed that baffles with more holes on the surface were more effective in reducing longitudinal forces, but baffles with a single central cavity could reduce vertical forces, and for controlling moments, single-cavity baffles proved to be more effective. Yg A et al. [24] conducted a numerical study on the influence of baffles on liquid sloshing in three-dimensional rectangular liquid tanks based on the nonlinear boundary element method, and the results showed that the shape of the baffle plays a non-negligible role in the sloshing of liquid tanks.

Liang Lihua (2021) et al. [25] analyzed the influence of a T-baffle arrangement on the intrinsic period of the tank through computational fluid dynamics software, and the results showed that changing the number of T-baffles can increase the intrinsic period of the tank to a certain extent. Zhao Minghan (2022) [26] used the VOF method to numerically simulate liquid sloshing in a trapezoidal liquid tank in a vehicle, and added different longitudinal baffles in the liquid tank, and the results showed that the more damping baffles, the faster the stabilization of the liquid sloshing free liquid level, and the lower the bulkhead pressure. Dongxi Liu et al. [27] conducted a series of model experiments in a fully filled cylindrical tank containing two immiscible liquids. It was found that the separation surface rotary sloshing in a two-layer liquid system was much more intricate than one-layer liquid rotary sloshing due to the generation of multitudinous short waves in the long wave. In Changle Zhang et al. [28], based on an improved moving-particle semi-implicit method, the BM-MPS method, the damping effect of a vertical slotted screen under rotation excitation was simulated and studied, and the influence of baffle porosity and the rotation amplitude on the resonance period and impact pressure was discussed. The results showed that the porosity had an obvious effect on the resonance period. Qiong Zhang et al. [29] studied the sloshing characteristics of a liquid cargo compartment under combined rollover and surge excitation conditions. Sarat Chandra Mohapatra et al. [30] analyzed wave-induced forces and moments acting on a cylinder, along with the circumferential pressure distribution around a vertical cylindrical structure, using CFD simulations.

To sum up, in traditional marine engineering, the addition of baffles has been widely confirmed as an effective means of sloshing suppression. Based on this mature method, this study analyzes the problem of liquid tank sloshing in maritime satellite launch platforms.

This study focuses on the specific application scenario of offshore satellite launch platforms, where propellant storage tanks (usually storing liquid oxygen, kerosene, etc.) are fundamentally different from traditional ship ballast tanks or FPSO oil storage tanks. The propellant storage tanks used to serve the rocket launch process are extremely stable—structural resonances or propellant cavitation caused by sloshing loads can lead to catastrophic consequences (e.g., explosions). At the same time, the Sea Launch platform is extremely sensitive to space and weight. While sloshing is often suppressed by adding baffles in traditional offshore engineering, the applicability of large-sized baffles in a launch platform environment can take up valuable propellant space and potentially interfere with the filling/discharge process. With the rapid advancement of computer technology, numerical simulation has become one of the most important tools in the study of liquid sloshing. Therefore, this paper uses STAR-CCM+2020 version to take the cylindrical storage tank of the hypothetical offshore rocket satellite launch platform as the research object, compares and verifies the cylindrical model (the model is idealized), and then analyzes the inhibition effect of the damping plate on the sloshing in the cylinder liquid tank, and then analyzes the influence of different positions of the damping plate and different types of damping plates on the amplitude of the liquid tank sloshing and the bulkhead pressure, and studies their inhibition effects.

## 2. Numerical Analysis Methods and Theory

### 2.1. Governing Equations

The turbulence model selected in this study incorporates the continuity equation, energy equation, and momentum equation that satisfy Newton's second law. The continuity equation, based on the principle of mass conservation, is expressed as (1): the increase in mass within a control volume per unit time equals the net mass inflow into the volume during the same time interval.

$$\frac{dm}{dt} = \frac{d}{dt}\iiint_{V(t)} \rho dV = 0 \tag{1}$$

In the equation, *V(t)* represents the volume of the fluid element; $m$ denotes the mass of the fluid within the control volume; and $\rho$ stands for the fluid density (in this study, it is assumed that the fluid density remains constant during the modeling process).

According to the mass transfer Equation (2), Equation (3) can be derived.

Based on the mass transfer Equation (2), defining $\phi = \rho$ (mass density) leads to Equation (3).

$$\frac{d}{dt}\iiint_{V(t)} \phi dV = \iiint_{V(t)} [\frac{d\phi}{dt} + \phi\nabla \cdot \boldsymbol{v}]dV = \iiint_{V(t)} [\frac{\partial\phi}{\partial t} + \nabla \cdot (\phi\boldsymbol{v})]dV = \iiint_{V(t)} \frac{d\phi}{dt}dV + \iint_{S} \phi v_n dS \tag{2}$$

$$\frac{d}{dt}\iiint_{V(t)} \rho dV = \iiint_{V(t)} [\frac{d\rho}{dt} + \rho\nabla \cdot \boldsymbol{v}]dV = \iiint_{V(t)} [\frac{\partial\rho}{\partial t} + \nabla \cdot (\rho\boldsymbol{v})]dV = 0 \tag{3}$$

The continuity equation can be derived from the arbitrariness of the fluid element volume *V(t)*, as shown in (4):

$$\frac{d\rho}{dt} + \rho\nabla \cdot \boldsymbol{v} = \frac{\partial\rho}{\partial t} + \nabla \cdot (\rho\boldsymbol{v}) = 0 \tag{4}$$

As shown in Equation (5), the rate of change of fluid momentum with respect to time within a fluid element is equal to the force acting on the fluid element.

$$\frac{d}{dt}\iiint_{V} \rho\boldsymbol{v} dV = \iiint_{V} \rho\boldsymbol{F} dV + \iint_{S} \tau_n dS \tag{5}$$

In the equation, $S$ represents the boundary condition; $\boldsymbol{F}$ denotes the body force; $\tau_n$ stands for the surface pressure.

The material derivative formula is given by (6):

$$\frac{d}{dt}\iiint\limits_{V} \boldsymbol{a}dV = \iiint\limits_{V} [\frac{d\boldsymbol{a}}{dt} + \boldsymbol{a}\nabla \cdot \boldsymbol{v}]dV = \iiint\limits_{V} \frac{d\boldsymbol{a}}{dt}dV + \iint\limits_{S} v_n \boldsymbol{a} dS \tag{6}$$

where $a$ is an arbitrary vector.

Application of the material derivative Formula (6):

Substitute the integrand $\boldsymbol{a} = \rho \boldsymbol{v}$ (momentum density vector) into (6):

$$\frac{d}{dt}\iiint\limits_{V} \rho \boldsymbol{v} dV = \iiint\limits_{V} \frac{d(\rho \boldsymbol{v})}{dt}dV + \iint\limits_{S} (\rho \boldsymbol{v}) v_n dS \tag{7}$$

where $v_n = \boldsymbol{v} \cdot \boldsymbol{n}$ denotes the velocity component normal to the boundary surface.

Combine Equations (5) and (7):

Substituting the left-hand side of (5) with expansion (7), the integral form of the momentum equation can be derived as

$$\iiint\limits_{V} \frac{\partial(\rho \boldsymbol{v})}{\partial t}dV + \iint\limits_{S} \rho v_n \boldsymbol{v} dS = \iiint\limits_{V} \rho F dV + \iint\limits_{S} \tau_n dS \tag{8}$$

### 2.2. Turbulence Model

Based on computational fluid dynamics (CFD) software (Siemens Star- CCM+ 2020 version), this paper analyzes and studies the liquid sloshing effects in baffled tanks and the motion performance of marine platforms equipped with these tanks. The finite volume method (FVM) is employed to solve the governing equations. The K-Epsilon turbulence model is adopted for the calculations. This K-Epsilon turbulence model includes transport equations for the turbulent kinetic energy [31] and turbulent dissipation rate, and the turbulent eddy viscosity is obtained by solving these equations.

The transport equations for the kinetic energy and turbulent dissipation rate are as follows:

$$\frac{\partial(\rho k)}{\partial t} + \nabla \cdot (\rho \mathbf{U} k) = \nabla \cdot \left( \left( \mu + \frac{\mu_t}{\sigma_k} \right) \nabla k \right) + P_k - \rho \varepsilon \tag{9}$$

$$\frac{\partial(\rho \varepsilon)}{\partial t} + \nabla \cdot (\rho \mathbf{U} \varepsilon) = \nabla \cdot \left( \left( \mu + \frac{\mu_t}{\sigma_k} \right) \nabla \varepsilon \right) + C_{1\varepsilon} P_k \frac{\varepsilon}{k} - C_{2\varepsilon} \rho \frac{\varepsilon^2}{k} \tag{10}$$

where $k$ is the transport equation for energy and $\varepsilon$ is the transport equation for the turbulent dissipation rate.

### 2.3. Boundary Conditions and Initial Conditions

Inside the liquid tank, the fluid satisfies the continuity equation, while the portion occupied by the liquid varies with time. The tank is not entirely filled with liquid, and the boundary conditions within the tank primarily consist of the wall conditions and the free surface.

(1) Wall boundary condition:

At wall boundaries, the no-slip condition and the slip condition are applied to the fluid. No-slip condition (the fluid exhibits no relative slip motion at the solid wall boundary):

$$\boldsymbol{v} = \boldsymbol{v}_b \tag{11}$$

Slip condition (fluid normal velocity equals wall velocity at the boundary, with unrestricted tangential slip velocity):

$$v \cdot n_b = v_b \cdot n_b \tag{12}$$

Here, $v_b$ represents the velocity of the liquid tank in the relative coordinate system. If the tank remains undeformed, then $v_b = 0$.

(2) Free-surface condition

At the free surface, the liquid satisfies both kinematic and dynamic conditions, and the water particles do not separate from the free surface. Additionally, this study neglects the surface tension of the liquid. Therefore, the boundary condition at the free surface is

$$\begin{cases} \frac{\partial u_n}{\partial t} + \frac{\partial u_t}{\partial n} = 0 \\ -p + 2\mu \frac{\partial u_n}{\partial n} = -P_0 \end{cases} \tag{13}$$

where $-P_0$ is the surface atmospheric pressure, $v$ is the fluid velocity, $u_n$ is the normal velocity of the free surface, and $u_t$ is the tangential velocity of the free surface.

This section establishes a systematic numerical framework based on hydrodynamic theory, presenting the theoretical foundations and computational methodologies involved in this study. The developed framework lays a solid foundation for subsequent research.

## 3. Validation and Analysis of the Numerical Model of Tank Sloshing

### 3.1. Model Creation and Calculation Settings

In order to verify the accuracy of the model and numerical method in this paper, the natural frequency of the liquid tank swing in the cylindrical container was numerically simulated, and the numerical results were compared with results from the literature and theoretical values. The three-dimensional cylindrical liquid tank in Pan Lijian (2007) [32] was selected for the model, and the dimensions of the cylindrical container are shown in Table 1: the diameter of the container was $D$ = 0.6 m, the radius was $R$ = 0.3 m, the height $H$ = 0.6 m, the thickness of the container wall was $d$ = 0.01 m, the medium inside the container was water, and the free liquid-level height was $h$ = 0.3 m when the liquid was stationary. The turbulent flow model is used for the computational model, the VOF model is used for multiphase flow, the implicit unsteady solver is used for the iterative method, and the second-order time discrete format is selected. The top is set to the pressure outlet, and the rest of the faces are all set to non-slip walls. A model of the cylindrical liquid tank and the free liquid-level contour of the liquid at rest are shown in Figures 1 and 2.

**Table 1.** Dimensions of the cylindrical tank model.

| Diameter (m) | Height (m) | Wall Thickness (m) | Free Liquid-Level Height |
|---|---|---|---|
| 0.6 | 0.6 | 0.01 | 0.3 |

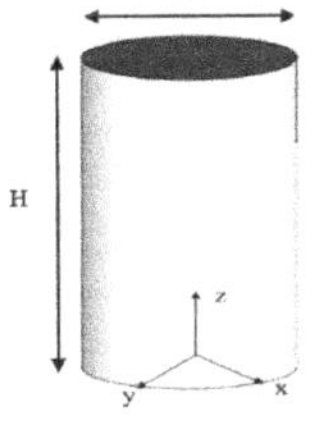


**Figure 1.** Cylindrical fluid tank model.

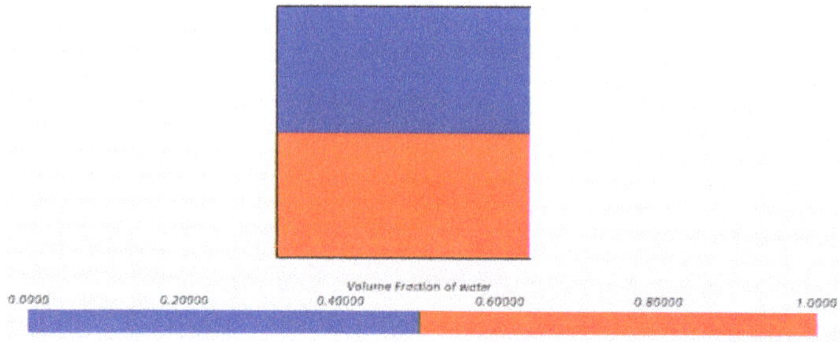


**Figure 2.** Still free-surface cloud map.

In this example, the meshing submodel of the surface reconstruction and cutting body mesh generator in STAR-CCM+ are adopted, and two layers of volume control are used to encrypt when the free liquid surface is shaking and when the liquid surface is close to stationary to ensure accurate capture of the waves; the surface reconstruction can optimize the surface shape of the liquid tank and facilitate the generation of volume meshes, as shown in Figure 3.

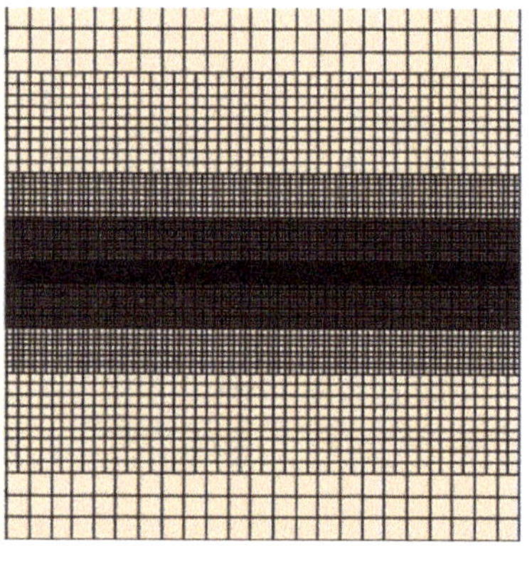

**Figure 3.** Schematic diagram of mesh (X-Z plane).

### 3.2. Grid Convergence Analysis

In this section, the grid is divided into 20 grids in the direction of the wave height of the free liquid surface, and the grid density encryption is carried out at two free liquid surfaces; the three grid sizes of 0.025 m, 0.027 m, and 0.03 m are analyzed for convergence analysis; the specific size information of the three grid schemes is shown in Table 2.

**Table 2.** Grid size settings.

| Grid | Minimum Size (m) | Maximum Size (m) | Number of Units (N) |
|---|---|---|---|
| M1 | 0.0016 | 0.025 | 2,945,872 |
| M2 | 0.0017 | 0.027 | 2,090,696 |
| M3 | 0.0019 | 0.03 | 1,645,850 |

Figure 4 shows the time history curves of the wave height at the left bulkhead of the cylinder fluid tank for the three different grid sizes. As can be seen from the figure, the meshes M2 and M3 can still converge in the initial stage (10–20 s) relative to the fine mesh M1. There is little difference between the wave simulation results of the grid schemes M1 and M2, and in general, the numerical results under the three grid schemes are approximately consistent.

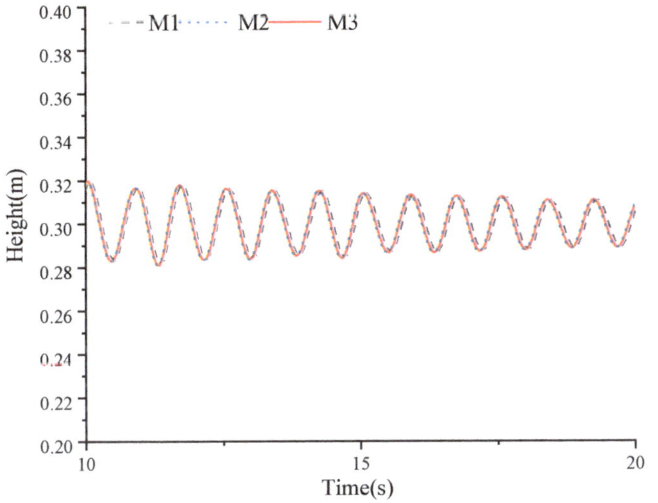


**Figure 4.** The wave elevation curve of left bulkhead for different grid sizes (10–20 s).

Table 3 shows an analysis of the average peak error of the left bulkhead wave height under the different grid schemes. As can be seen from the table, the relative error of both grids M2 and M3 does not exceed 5% relative to the fine grid M1. The results of the grid convergence analysis show that the liquid sloshing amplitude of the STAR-CCM simulated cylindrical liquid tank model is not sensitive to the grid size, and the numerical simulation results under the three grid sizes are very reliable. In order to ensure the accuracy of the numerical calculations and account for the limitation of computing resources, the M2 grid scheme was selected for the meshing of the liquid tank.

**Table 3.** The average peak of left bulkhead wave height for different grid sizes.

| Grid Scheme | M1 | M2 | M3 |
|---|---|---|---|
| Average peak (m) | 0.3149 | 0.3133 | 0.3070 |
| Relative error (%) | 0 | −0.5 | −2.51 |

### 3.3. Time-Step Convergence Analysis

The choice of time-step size determines the level of calculation accuracy and the consumption of computing resources, and grid scheme M2 was analyzed and selected in this section to carry out a study of the influence of time-step size on the convergence of the calculation results; a total of three time steps were set, which are 0.003 s, 0.0035 s, and 0.004 s; the curves of the wave heights for the three different time steps with the time step are shown in Figure 5.

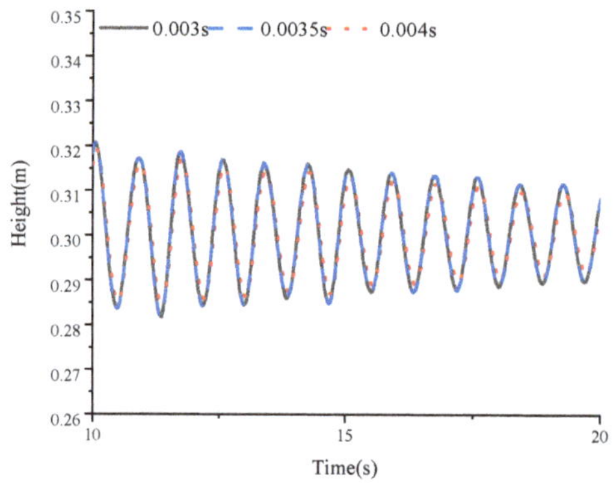


**Figure 5.** The wave elevation curve of left bulkhead for different time steps (10–20 s).

According to the information analysis in Table 4, it can be seen that the wave surface at the left bulkhead of the liquid tank gradually stabilizes and attenuates, and the average amplitude error of the two is 0.03% when the numerical calculation time step is 0.035 s and 0.003 s, and the relative error is slightly larger, at 0.38%, when the calculation time step is 0.004, and the errors of the three calculation results are within the allowable range, indicating that the numerical calculation models of the three time steps have reached convergence. Considering the stability of the calculation results and the calculation cost, the calculation time step of 0.0035 s was chosen.

**Table 4.** Average peak of left bulkhead wave height for different time steps.

| Time-Step Schemes | Average Peak (m) | Relative Error (%) |
|---|---|---|
| 0.003 | 0.3146 | 0 |
| 0.0035 | 0.3145 | −0.03 |
| 0.004 | 0.3133 | 0.38 |

### *3.4. Verification and Analysis of the Accuracy of the Numerical Model of Liquid Tank Sloshing*

The natural frequency of the tank does not change with the external conditions, it is an inherent property of the tank, and it is only related to the shape and depth of the tank. Please check meaning retained In order to verify the accuracy of the numerical method of simulating liquid tank sloshing in this paper, the numerical calculation results in this paper are compared with the theoretical values of the empirical formula of the natural frequency of a partially filled cylinder liquid tank derived from Pan Lijian's study and Franklin T Dodge (2000) [32,33], so as to verify the accuracy of the numerical calculation results in this paper. The formula for calculating the natural frequency of the cylinder fluid tank is as follows:

$$\omega_n^2 = \frac{1.84g}{R}\tanh(1.84h/R), \tag{14}$$

where $h$ is the depth of the liquid at rest, $R$ is the radius of the cylinder, $n$ is the order of the wave, and $g$ is the acceleration due to gravity. $\omega$ is the circular frequency; when $n$ is 1, it corresponds to the first-order natural frequency, and when $n$ is 2, it corresponds to the second-order natural frequency.

When a partially filled container is disturbed by an external force or is subjected to a transient horizontal excitation at the initial moment, the liquid in the tank shakes. For example, the tank suddenly accelerates or gives the liquid in the tank an initial wave shape at the initial moment, allowing it to swing freely under gravity. In this section, the natural frequencies of the liquid in a three-dimensional cylindrical fluid chamber are verified by numerical simulation, setting the initial wave surface to the first-order mode shape; the initial wave surface parameters are shown in Table 5.

**Table 5.** Initial wave surface parameter.

| Case | $\eta_0$ | Wave Surface Function | STAR-CCM+ Volume Fraction Setting |
|---|---|---|---|
| First-order mode shapes | 0.03 | $\eta = \eta_0 \sin(\pi x/L)$ | `($${Position}[2] <= (0.3 − 0.03 * sin(3.14 × $${Position}[0]/0.6)))?1:0` |

In Table 5, 0.03 is the amplitude, x is the position coordinate along the length, and $L$ is the length in the x-direction.

The three-dimensional cylinder liquid tank sloshing simulation was carried out by using the STAR-CCM+ software, and the initial conditional volume fraction field function was set by the STAR-CCM+ user-defined field function, and a schematic diagram of the initial free liquid level, as shown in Figure 6, was obtained. The initial wave surface moves freely, attenuating at its natural frequency under the action of gravity; a high free attenuation curve at the left bulkhead wave under the condition of the initial first-order mode shape wave surface is obtained, as shown in Figure 7, and the pressure change curve at the center point of the left bulkhead is shown in Figure 8.

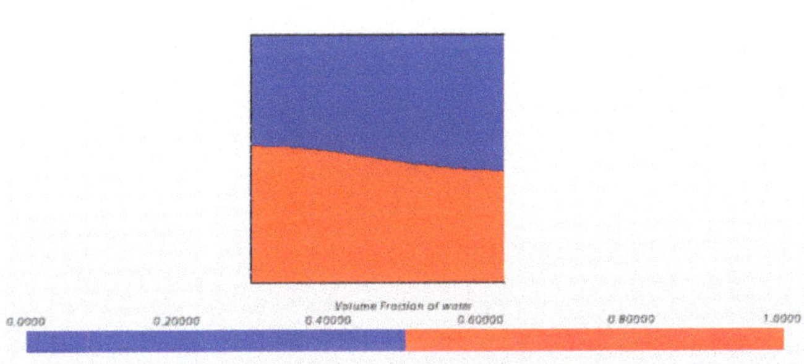


**Figure 6.** Visualization cloud diagram of free surface.

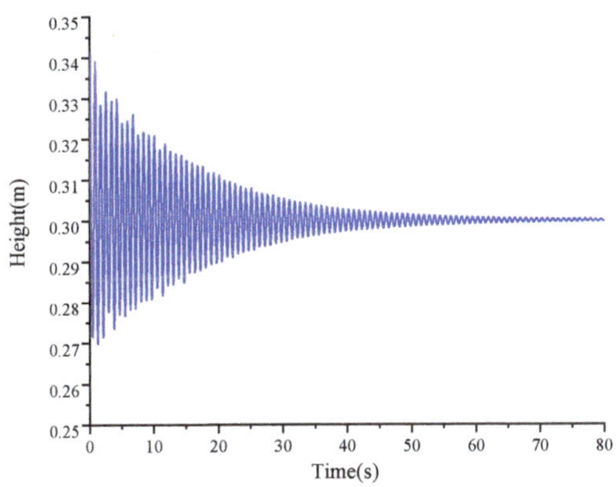


**Figure 7.** The wave elevation curve of left bulkhead.

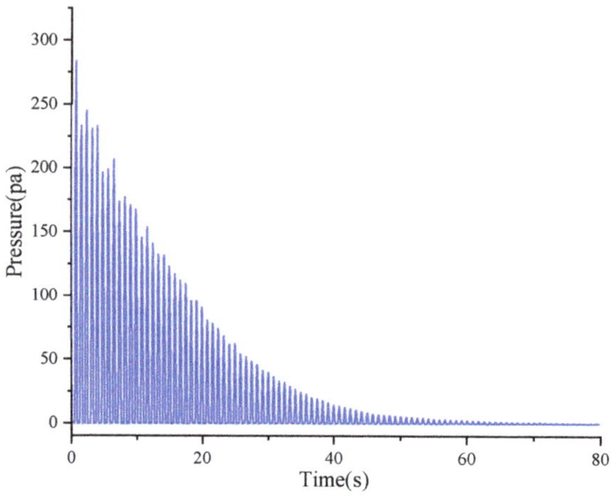


**Figure 8.** Pressure curve at center point of left bulkhead.

A Fourier transform analysis was carried out on the elapsed curve of the wave height at the left bulkhead, and then according to the spectral density formula, the energy spectrum shown in Figure 9 was obtained. The formula for spectral density is

$$S(\omega) = \left(A^2/2\right)/d\omega, \tag{15}$$

where $A$ is the amplitude value obtained by the Fourier transform.

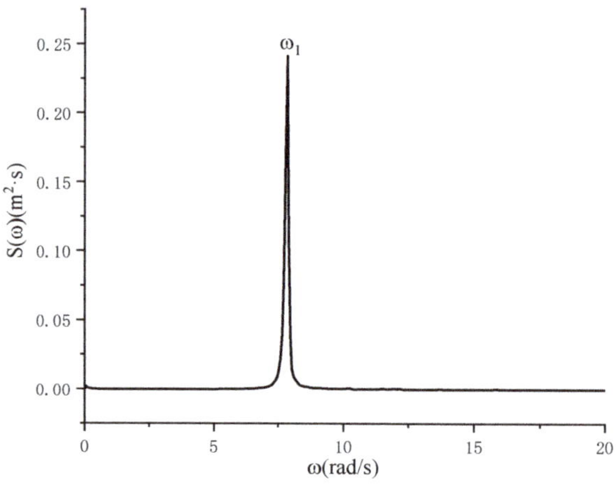


**Figure 9.** Energy spectrum.

It is evident from the graph that there is a distinct peak in the energy spectrum, which corresponds to a first-order natural frequency of 7.56 rad/s. The numerical results are compared with the theoretical values obtained by the empirical formula and the median values in the relevant literature, and the error results obtained by the analysis are shown in Table 5. The formula for calculating the error of the natural frequency is

$$\Delta\omega_n = \frac{\omega_{theory} - \omega_{STAR-CCM+}}{\omega_{theory}} \times 100\%, \tag{16}$$

According to the analysis in Table 6, the error between the calculated results and the theoretical values in this paper is 3.83%, while there is almost no error compared to the calculation results in the related literature, indicating that the numerical model can simulate the first-order wave surface sway problem in the cylindrical fluid tank well, and at the same time can maintain good accuracy, which provides strong support for the following research on the anti-sway effect of the damping plate in the cylindrical fluid tank.

**Table 6.** Relative error analysis of natural frequency.

| | Numerical Result | Theoretical Value | Literature Calculation |
|---|---|---|---|
| $\omega_n$(rad/s) | 7.852 | 7.562 | 7.853 |
| Error (%) | | 3.83 | 0.01 |

## 4. Analysis of the Effect of the Damping Plate on Suppressing Sloshing in the Liquid Tank

The numerical simulation of the cylindrical liquid tank has been verified above, and this subsection takes the hypothetical cylindrical liquid tank in the column of the maritime satellite launch platform as the research object. As shown in Table 7, the diameter of the cylindrical liquid tank D1 is 10.18 m, the height of the cylindrical liquid tank $H_1$ is 10.9 m, the water filling volume in the liquid tank is 50%, and the free liquid level height is 5.45 m when stationary. In order to study the inhibition effect of the damping plate on the wave surface sloshing in the cylinder liquid tank, an open-hole damping plate with a thickness of 0.03 m was designed: the diameter of the damping plate was the same as that of the cylinder, the opening position was located at the center of the circle, the aperture was $r$ = 5.09 m, and the damping plate was one fifth of the height of the stationary free liquid surface. As shown in Figure 10, the model compares the wave surface sloshing of the liquid tank with and without damping plates to verify the inhibition effect of the damping plates on the sloshing in the liquid tank.

**Table 7.** Dimensions of the cylindrical tank model with baffles.

| $D_1$ (m) | $H_1$ (m) | $r$ (m) |
|---|---|---|
| 10.18 | 10.9 | 5.09 |

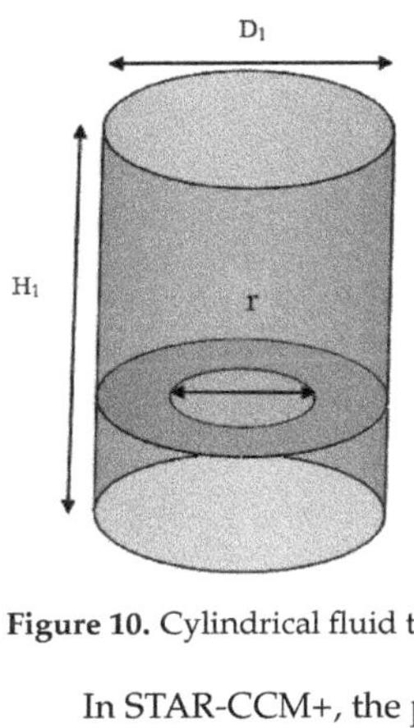


**Figure 10.** Cylindrical fluid tank model with baffle.

In STAR-CCM+, the physical model is defined based on the physical continuum. For a 3D model, 3D is first selected. The sloshing of the liquid in the tank is an unsteady flow, so the implicit unsteady time solver is chosen. Since there are two fluids in the tank, water and air, the Eulerian multiphase flow is selected to define the fluid, and the fluid domain volume model is further selected. For viscous fluids, the choice of laminar and turbulent flow depends on the Reynolds number. In this example, a turbulence model is used for the simulation, and gravity is selected to generate traveling waves. The pressure near the free surface in the initial conditions is set at a standard atmospheric pressure of 101,325 Pa.

Based on the time step of 0.0035 s chosen in Section 3.3 and the meshing method chosen in Section 3.2, this subsection uses the meshing submodel of surface reconstruction and cutting body mesh generator in STAR-CCM+, and uses three-layer volume control to encrypt the vicinity of the free liquid surface sloshing, the stationary place of the free liquid level, and the vicinity of the damping plate to ensure the accuracy of wave capture, and the basic mesh size of the 3D column liquid tank model is set to 0.25 m. The meshing is shown in Figure 11.

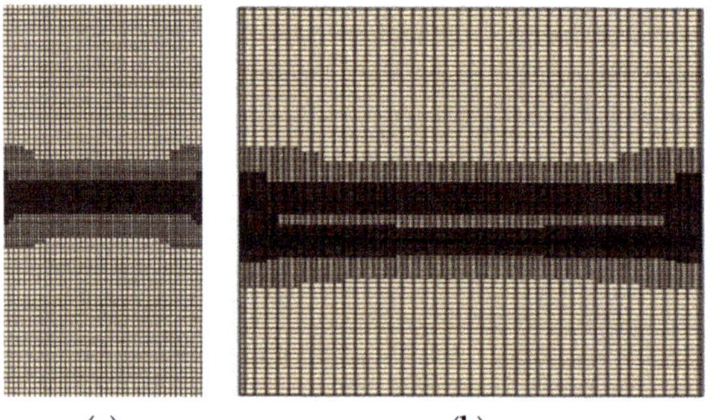

(a) (b)

**Figure 11.** Schematic diagram of mesh (X-Z plane). (**a**) No damping plates; (**b**) With plate damping.

This section still initiates the fluid motion by defining the initial wave surface within the tank via a newly created field function at time zero, thereby enabling observation of the free sloshing damping characteristics. When the liquid in the liquid tank of the three-dimensional cylindrical column is at rest, the free liquid level is half of the height of the column, that is, 5.45 m, and the liquid filling rate is 50%; the volume fraction scalar contour is used to record the shaking of the liquid in the liquid tank, as shown in Figure 12.

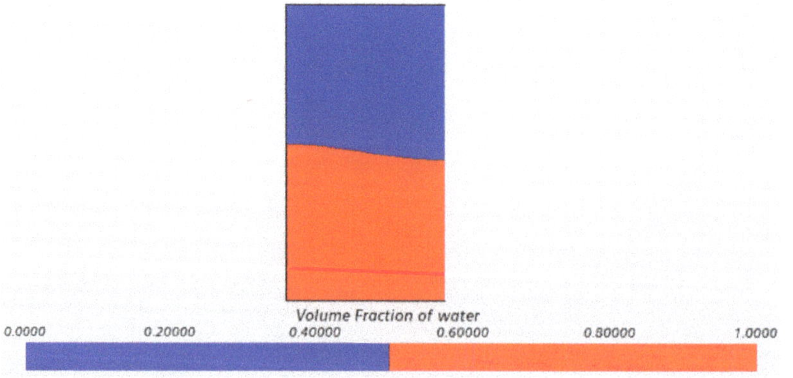


**Figure 12.** Visual cloud map.

Figure 13 is the diachronic curve of the wave height change at the left bulkhead with and without a damping plate, and the analysis of the information in the diagram shows that the liquid-level height in the liquid tank gradually decays with time, and the liquid-level height attenuation is particularly obvious after adding the damping plate. After 60–100 s of wave surface attenuation, the average amplitude of the liquid surface without the damping plate is 0.941 m, and the average amplitude of the liquid surface during this period is 0.097 m after adding the damping plate, and the wave surface attenuation amplitude can reach 89.69%, indicating that the damping plate can effectively suppress the liquid shaking in the cylindrical liquid tank, and at the same time make the wave surface reach a calm state quickly. The reason for this analysis is mainly due to the interference effect of the damping plate on the movement of the water quality point of the wave surface, which accelerates the energy consumption of the wave surface, and the addition of the damping plate also changes the natural frequency of the liquid tank. A Fourier transform was carried out on the diachronic curve of the wave height change at the left bulkhead of the liquid tank with or without a damping plate; from the energy spectrum obtained by the Fourier transform on the wave height change curve at the left bulkhead in the case of the damping plate, as shown in Figures 14 and 15, it can be seen that the natural frequency of the first-order sloshing of the wave in the liquid tank without a damping plate is 1.926 rad/s, and the natural frequency of the first-order sloshing is 1.643 rad/s after adding the damping plate. This shows that the addition of the damping plate can change the natural frequency of the liquid sloshing in the liquid tank, reduce its movement, and achieve an inhibition effect; and the main reason for the reduction in the shaking amplitude of the damping plate is the increase of physical dissipation.

Figure 16 is the pressure change curve of the center point of the left bulkhead with and without the damping plate, and the information in the diagram shows that the pressure on the left bulkhead in the liquid tank gradually decreases with the time after the beginning of liquid sloshing. After 40–80 s, the maximum pressure on the left bulkhead without the damping plate is 4837.96 Pa, and the maximum pressure on the left bulkhead with the damping plate is 1104.20 Pa; it can be clearly seen that the pressure on the left bulkhead is significantly reduced after adding the damping plate, which is due to the addition of the damping plate suppressing the shaking of the wave surface, thereby reducing the impact force on the bulkhead and improving the stability.

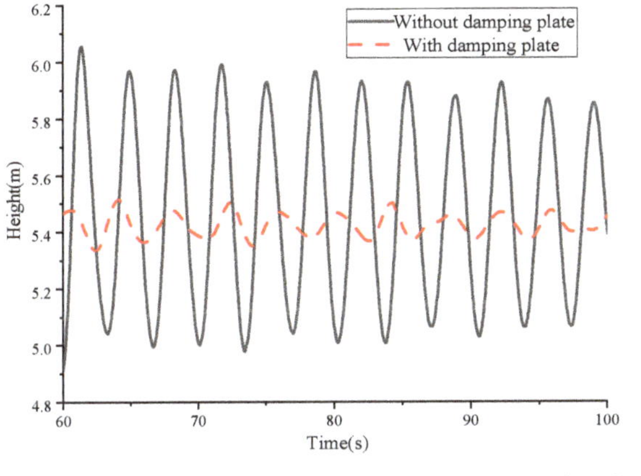


**Figure 13.** The wave elevation curve of left bulkhead with and without baffle.

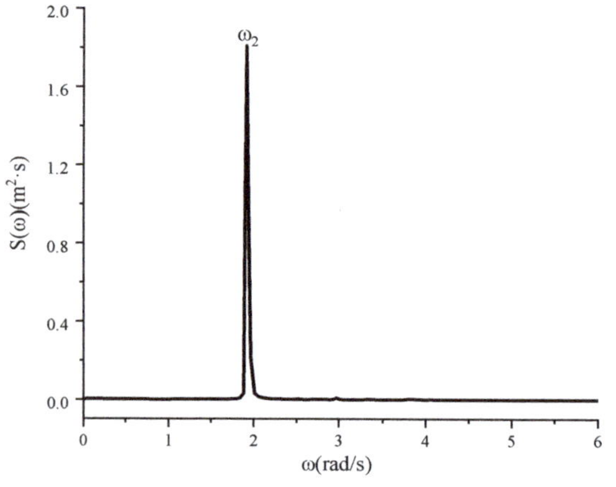


**Figure 14.** The energy spectrum of left bulkhead wave height variation duration curve without baffle.

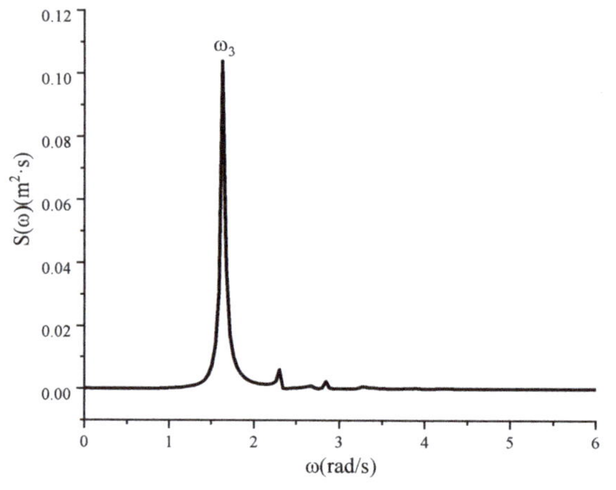


**Figure 15.** The energy spectrum of left bulkhead wave height variation duration curve with baffle.

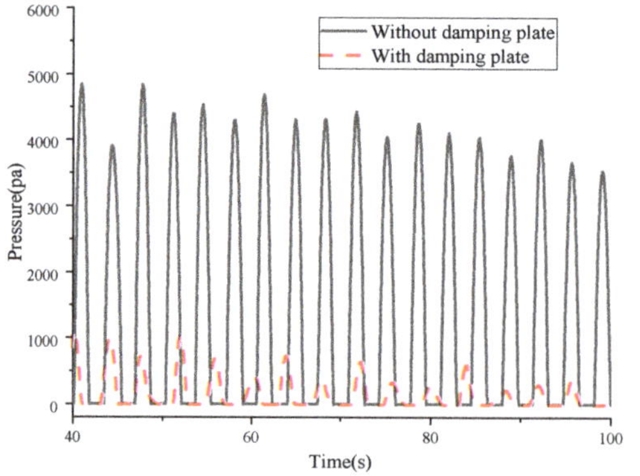


**Figure 16.** Pressure curve at center point of left bulkhead with and without baffle.

## 5. The Influence of the Position of the Damping Plate on the Sloshing in the Liquid Tank

### 5.1. Damping Separator Parameters

In order to study the inhibition of sloshing in the liquid tank when the damping plate is placed at different heights, the inhibition effect of the damping plate at three different positions in the liquid tank was numerically calculated. The initial static water-level height of the liquid tank is $H_2$, the height of the upper surface of the damping plate from the stationary water surface is $h_1$, and the design damping partition is located at $h_1/H_2 = 1/10$, $h_1/H_2 = 1/5$, $h_1/H_2 = 2/5$; the specific position layout of the damping plate is shown in Figure 17.

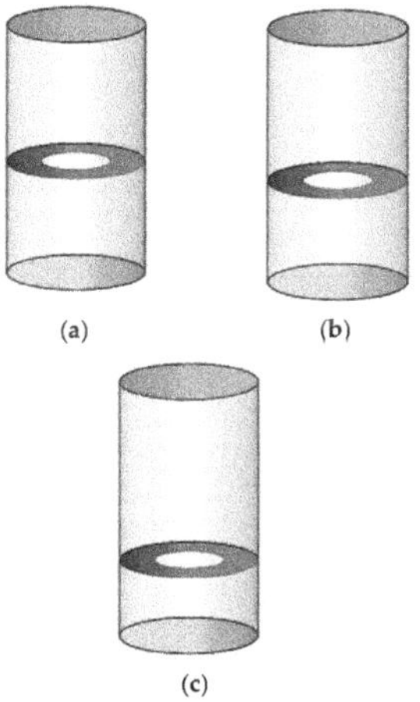


**Figure 17.** Layouts of different baffles. (**a**) $h_1/H_2 = 1/10$; (**b**) $h_1/H_2 = 1/5$; (**c**) $h_1/H_2 = 2/5$.

### 5.2. The Influence of Damping Plates in Different Positions on the Amplitude and Pressure in the Liquid Tank

In order to analyze the suppression effect of the damping plate on the wave surface sloshing in the liquid tank, this subsection compares the wave height elapsed curves and the pressure at the left bulkhead in the liquid tank for three different damping plate positions, and then obtains the average amplitude and maximum pressure in the liquid tank at three different damping plate positions, so as to analyze the difference in the damping plate position on the sway suppression effect in the liquid tank.

Figure 18 is the elapsed curve of the wave height change at the left bulkhead under different damping plate positions for 40–100 s; the information in the analysis diagram shows that as time passes, the liquid-level height in the liquid tank gradually attenuates, and at the same time, when the damping plate is away from the liquid-level heights $h_1/H_2 = 1/10$ and $h_1/H_2 = 1/5$, the free liquid level attenuation is obvious: the wave amplitude is small and close to stable. From the graph analysis, it can be seen that when $h_1/H_2 = 1/10$, the average wave amplitude is 0.0497 m, and when $h_1/H_2 = 1/5$, the average amplitude is 0.0977, and the average amplitude difference is about 0.048 m, which indicates that when the damping plate is close to the liquid level, the closer the damping plate is to the liquid level, the higher its inhibition effect on the liquid sloshing in the liquid tank, but the lifting effect is smaller; when the damping plate is located at $h_1/H_2 = 2/5$, it is obvious from the figure that the wave surface amplitude is larger, compared with $h_1/H_2 = 1/5$ and $h_1/H_2 = 1/10$. The damping plate has a poor inhibitory effect on the liquid sloshing in the liquid tank, and the average amplitude is 0.5223 m. The reason for this is mainly due to the fact that when the liquid in the liquid tank is shaken by the first-order wave surface, the energy of the liquid shaking is mainly concentrated near the wave surface; when the damping plate is close to the wave surface, the wave surface can interact with the damping plate, consume the water quality point energy of the wave surface, and destroy the wave surface movement. Figures 19 and 20 show the first-order-shaking energy spectrum of the liquid tank wave surface when the damping plate is arranged at $h_1/H_2$ = 1/10 and $h_1/H_2$ = 2/5, and it can be seen from the figure that when the position of the damping plate is $h_1/H_2$ = 1/10, the natural frequency of the liquid surface sloshing in the tank is 1.341 rad/s, which is quite different from the natural frequency of the first-order wave surface sloshing without the damping plate, indicating that the wave surface motion has been significantly reduced, making it reach a stable state earlier. When the position of the damping plate is $h_1/H_2$ = 2/5, the natural frequency of the liquid-level swing in the liquid tank is 1.884 rad/s, which is less different from the natural frequency of the first-order wave surface swing without the damping plate, which indicates that the interference of the damping plate arranged at this position on the movement of the liquid surface is small, and the suppression effect is poor when the damping plate is far away from the wave surface.

Figure 21 is the curve of pressure change at the center point of the left bulkhead under different damping plate positions between 20 and 80 s, and the figure shows that as time passes the pressure of the left bulkhead of the liquid tank gradually decreases, which is similar to the phenomenon of liquid swing amplitude in the liquid tank, and when the damping plate is arranged at the height $h_1/H_2$ = 1/10, because the liquid level is close to stability, the pressure change amplitude at the left bulkhead caused by the wave surface is small, almost zero; when the damping plate arrangement height $h_1/H_2$ = 1/5, the left bulkhead is still subjected to a certain pressure because the liquid in the liquid tank still has a small-amplitude oscillation, and its maximum value is 2246.4 Pa; when the damping plate is arranged at the height $h_1/H_2$ = 2/5, the left bulkhead of the liquid tank is subjected to a large impact because the liquid tank shake is less disturbed by the damping plate, the wave surface sloshing attenuation is smaller, and the left bulkhead of the liquid tank is subjected to a larger impact. The maximum pressure is 3608.1 Pa, which indicates that the damping plate is arranged near the wave surface, which can effectively reduce the force at the bulkhead, reduce the impact on the liquid tank, and improve its stability and safety.

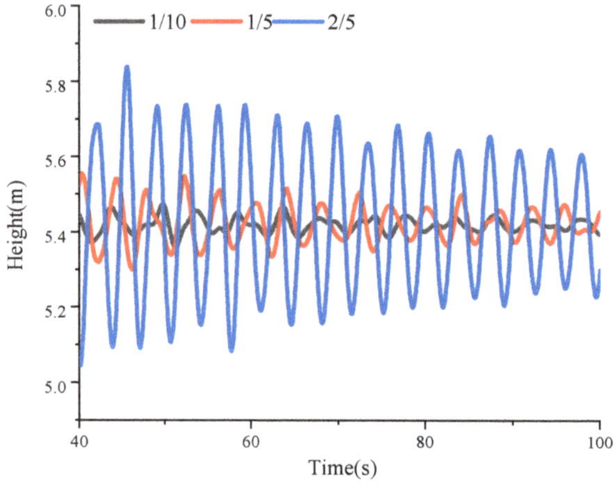


**Figure 18.** The wave elevation curve of left bulkhead at different baffle heights.

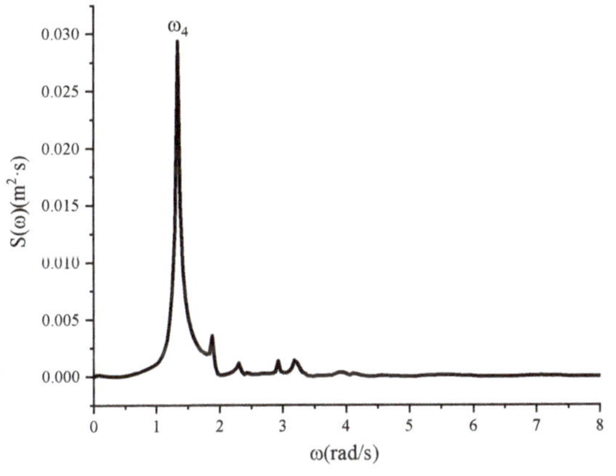


**Figure 19.** Energy spectrum of left bulkhead wave height variation duration curve ($h_1/H_2 = 1/10$).

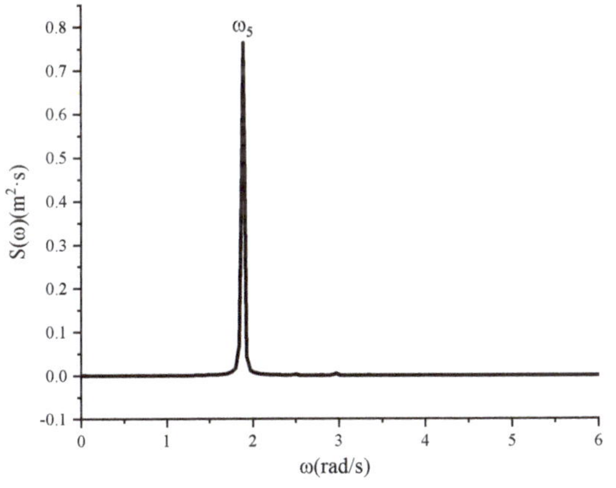


**Figure 20.** Energy spectrum of left bulkhead wave height variation duration curve ($h_1/H_2 = 2/5$).

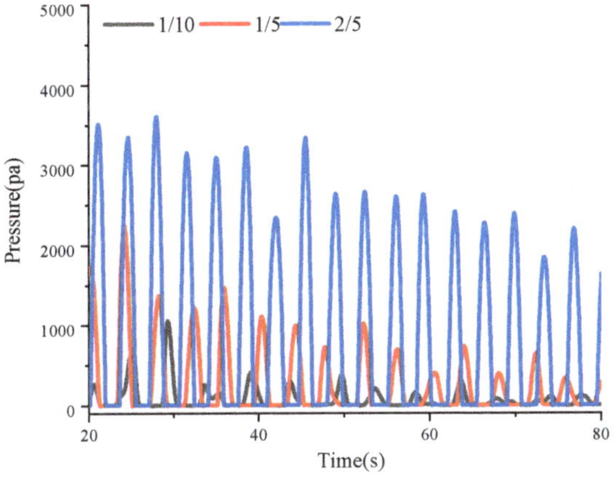


**Figure 21.** Pressure curve at center point of left bulkhead at different baffle heights.

# 6. The Influence of the Damping Plate Type on the Sloshing in the Liquid Tank

## 6.1. Damping Partition Parameters

From the above research, it can be seen that the closer the damping plate is to the free liquid level, the better the sloshing inhibition effect of the liquid tank. In this section, the opening type of the damping plate is studied, and the inhibition effect of damping plates with different numbers of openings on the sloshing of the liquid tank is analyzed while keeping the position of the separator one fifth meter away from the free liquid level. The damping plates are all circular and remain perpendicular to the column, the amount of solid material remaining in the cross-sectional area of the baffle plate is almost the same, about 75%; the numbers of openings in the damping plates are one, three, and six, respectively; and the layouts of the different types of damping plates are shown in Figure 22.

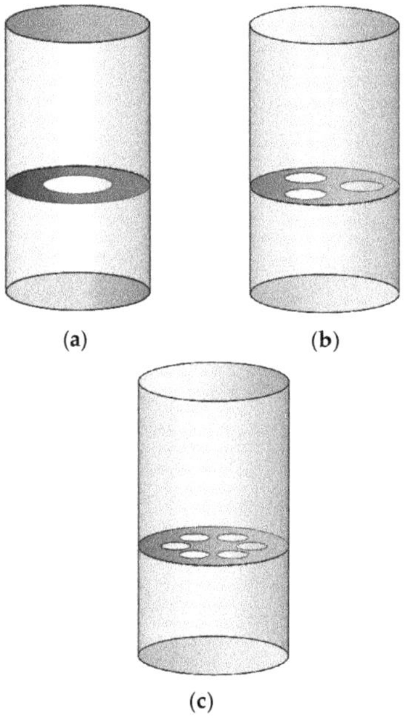


**Figure 22.** Layouts of different baffles. (**a**) 1 hole; (**b**) 3 holes; (**c**) 6 holes.

## 6.2. The Influence of Different Types of Damping Plates on the Amplitude and Pressure in the Liquid Tank

In order to further analyze the inhibition effect of different types of damping plates on the sloshing in the tank, this subsection compares the wave height elevation curves and the pressure on the left bulkhead for the three types of damping plates, and then obtains the average amplitude and maximum pressure in the tank with three different types of damping plates, so as to analyze the difference in the sloshing inhibition effect of different types of damping plates on the tank.

Figure 23 is the elapsed curve for the wave height at the left bulkhead of the liquid tank with different types of damping plates between 40 and 100 s; it shows that as time passes, the liquid-level height in the liquid tank also gradually attenuates, and at the same time, when the number of damping plate openings is three or six, the free liquid level attenuation amplitude is small, close to stable; and it can be seen from the analysis of the figure that when the number of holes in the damping plate is one, the average amplitude is 0.1465 m, and when the number of holes in the damping plate is three, the average amplitude is 0.0848 m, and the average amplitude difference is about 42.12%, indicating that when the opening rate of the damping plate is constant, the increase in the number of holes has a higher inhibitory effect on the liquid sloshing in the tank. When the number of openings of the damping plate increases to six, the average amplitude of the wave surface shaking is 0.0728 m; that is, for three more openings in the damping plate, the inhibition effect of the damping plate on the liquid sloshing in the liquid tank is not significantly improved. As shown in Figures 24 and 25, when the number of openings in the damping plate is three, the natural frequency of the liquid level swing in the liquid tank is 1.747 rad/s, and when the number of holes in the damping plate is six, the natural frequency of the liquid level swing in the tank is 1.759 rad/s, compared with the natural frequency of 1.643 rad/s when the number of openings is one. The increase in the number of holes changes the natural frequencies to a certain extent; however, there is almost no difference between the natural frequency of the first-order wave surface sloshing when the number of holes in the damping plate is three and that when the number of holes in the damping plate is six, which indicates that with an increase in the number of holes in the damping plate, the inhibition effect on the liquid sloshing in the liquid tank is strengthened, but when the number of holes increases to a certain number, the inhibition effect is barely enhanced, which further verifies the above conclusion.

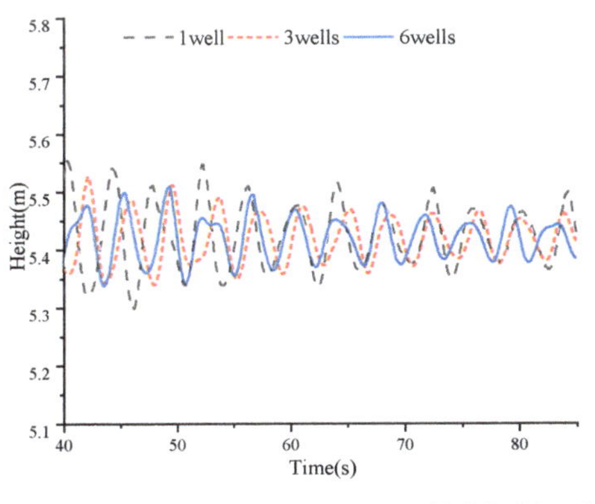


**Figure 23.** The wave elevation curve of left bulkhead for different types of baffle.

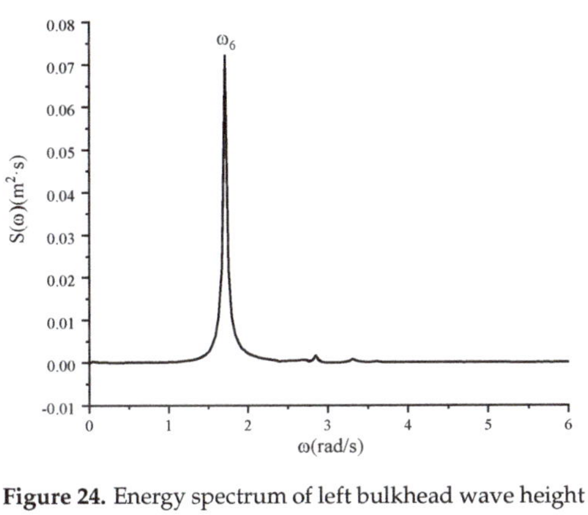


**Figure 24.** Energy spectrum of left bulkhead wave height variation duration curve (three open holes).

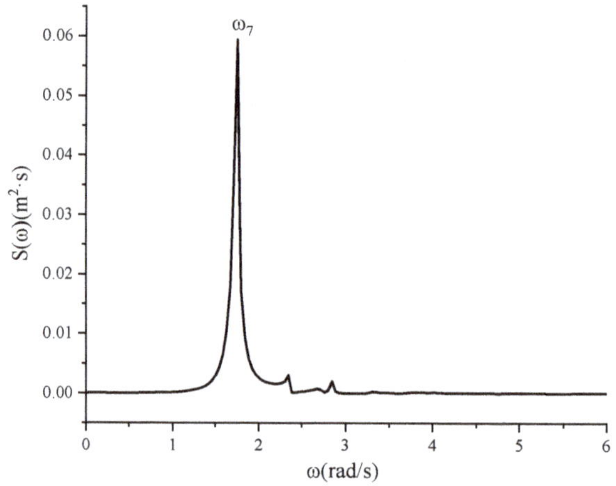


**Figure 25.** Energy spectrum of left bulkhead wave height variation duration curve (six open holes).

Figure 26 is the curve of pressure change at the center point of the left bulkhead of the liquid tank for different types of damping plates between 40 and 90 s; it shows that as time passes, the pressure on the left bulkhead of the liquid tank gradually decreases, which is similar to the phenomenon of the liquid swing amplitude in the liquid tank when the number of holes in the damping plate is one, because the liquid in the liquid tank still has a certain oscillation the left bulkhead is still subjected to a certain pressure, the maximum value of which is 1104.6 Pa. When the number of damping plate openings is three and six, the maximum pressure values are 853.60 Pa and 684.82 Pa, respectively, indicating that a greater number of damping plate openings can effectively reduce the force at the bulkhead, but the effect does not change significantly after reaching a certain number, and the number of damping plate openings needs to be reasonably set to reduce the impact of the liquid tank and improve its stability and safety.

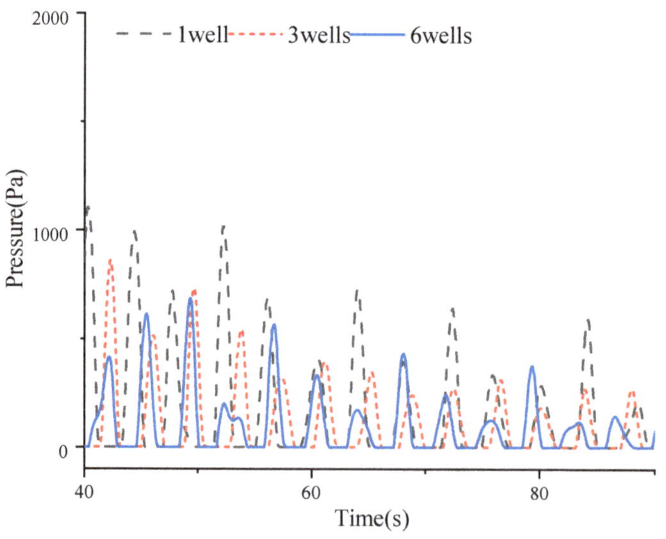


**Figure 26.** Pressure curve at center point of left bulkhead for different types of baffle.

## 7. Conclusions

First, this paper validates the numerical model based on a model in the literature. The reliability of the numerical simulation model of the cylindrical liquid tank in this paper is verified by comparing the calculated results of the numerical model with results in the literature and theoretical values.

Secondly, this paper takes the hypothetical liquid tank in the upright column of the maritime satellite launch platform (the model is the ideal model) as the research object, analyzes the free liquid surface movement of the liquid in the liquid tank with and without damping plates, and then sets up damping plates at different positions, as well as different types of damping plates, to analyze the influence of damping plates on liquid sloshing in the liquid tank, and draws the following conclusions:

(1) With time, the free liquid level in the liquid tank gradually decays and finally reaches a stationary state. After adding the damping plate, the natural frequency of liquid sloshing in the liquid tank is changed, and the attenuation of the free liquid level height decreases sharply, which can effectively inhibit the liquid surface movement in the cylindrical liquid tank, and at the same time, the wave surface can reach a stationary state quickly.

(2) In the case of the same type of damping plate, as the damping plate approaches closer to the free liquid level, the free liquid level shaking amplitude becomes smaller and smaller, and the pressure on the bulkhead is also significantly reduced, leading to a better inhibition effect on the wave surface movement of the liquid tank.

(3) With an increase in the number of openings in the damping plate, the shaking amplitude of the free liquid level gradually decreases, and the pressure on the bulkhead gradually decreases when the position of the damping plate remains unchanged. For increases beyond a certain number of openings in the damping plate, the inhibition effect on the shaking wave surface motion of the liquid tank does not significantly improve. With the damping plate position held constant, a three-perforation configuration yields peak sloshing suppression efficacy in the liquid tank. Simulations demonstrate a progressive reduction in free-surface wave amplitude and tank wall pressure with increasing perforation count. However, beyond a critical perforation density, further aperture additions provide diminishing returns in wave elevation motion control.

## 8. Limitations and Future Work

The research presented in this paper remains at a conceptual exploration stage, and its conclusions are primarily based on idealized model conditions. Consequently, the study on the offshore satellite launch platform is not yet sufficiently in-depth.

(1) This study is based on numerical simulation methods. Future research could conduct physical model tests to further validate and optimize the numerical analysis model presented here.

(2) During the design of the damping plate in this study, the variety of parameters optimized was limited. Future work could focus on optimizing the design of the damping plate.

(3) In this study, the launch platform model was simplified without considering factors such as the platform's superstructure or rocket launch parameters. Future research could incorporate considerations of actual engineering aspects of offshore launch platforms.

**Author Contributions:** Conceptualization, Y.P., Y.W., F.L. and G.X.; Methodology, Y.P., Y.W., F.L. and G.X.; Software, Y.P., Y.W., F.L. and G.X.; Validation, Y.P. and Y.W.; Formal analysis, Y.P. and Y.W.; Investigation, Y.P. and Y.W.; Resources, Y.P. and Y.W.; Data curation, Y.P. and Y.W.; Writing—original draft, Y.P. and Y.W.; Writing—review & editing, Y.P.; Visualization, Y.P. and Y.W.; Funding acquisition, Y.P. All authors have read and agreed to the published version of the manuscript.

**Funding:** This research received no external funding.

**Data Availability Statement:** The original contributions presented in this study are included in the article. Further inquiries can be directed to the corresponding author.

**Conflicts of Interest:** All authors state that the study was conducted without any commercial or financial relationship that could be interpreted as a potential conflict of interest.

## References

1. Park, J.J.; Kim, S.Y.; Kim, Y.; Seo, J.-H.; Jin, C.H.; Joh, K.H. Study on tank shape for sloshing assessment of LNG vessels under unrestricted filling operation. *J. Mar. Sci. Technol.* **2015**, *20*, 640–651. [CrossRef]
2. Yu, L.; Xue, M.; Zhu, A. Numerical Investigation of Sloshing in Rectangular Tank with Permeable Baffle. *J. Mar. Sci. Eng.* **2020**, *8*, 671. [CrossRef]
3. Golla, S.T.; Venkatesham, B. Experimental study on the effect of centrally positioned vertical baffles on sloshing noise in a rectangular tank. *Appl. Acoust.* **2021**, *176*, 107890. [CrossRef]
4. Ju, H.B.; Jang, B.S.; Ki-Ho, Y. Prediction of sloshing pressure and structural response of LNG CCS. *Ocean Eng.* **2022**, *266*, 112298. [CrossRef]
5. Mccarty, J.L.; Stephens, D.G. *Investigation of the Natural Frequencies of Fluids in Spherical and Cylindrical Tanks*; National Aeronautics and Space Administration: Washington, DC, USA, 1960.
6. Francescutto, A.; Armenio, V.; Rocca, L.M. On the roll motion of a ship with partially filled unbaffled and baffled tanks: Numerical and experimental investigation. In Proceedings of the Sixth International Offshore and Polar Engineering Conference, Los Angeles, CA, USA, 26–31 May 1996; Volume 3, pp. 377–386.
7. Gawad, A.; Ragab, S.A.; Nayfeh, A.H.; Mook, D.T. Roll stabilization by anti-roll passive tanks. *Ocean Eng.* **2001**, *28*, 457–469. [CrossRef]
8. Maleki, A.; Ziyaeifar, M. Sloshing damping in cylindrical liquid storage tanks with baffles. *J. Sound Vib.* **2008**, *311*, 372–385. [CrossRef]
9. Storey, J.M.; Poothokaran, J.; Kirk, D.R.; Kirk, D.R.; Gutierrez, H.; Schallhorn, P.A. Experimental, Numerical, and Analytical Slosh Dynamics of Water and Liquid Nitrogen in a Spherical Tank. In Proceedings of the AIAA/SAE/ASEE Joint Propulsion Conference, Salt Lake City, UT, USA, 25–27 July 2016.
10. Liu, J.; Zang, Q.; Ye, W.; Lin, G. High performence of sloshing problem in cylindrical tank with various barrels by isogeometric boundary element method. *Eng. Anal. Bound. Elem.* **2020**, *114*, 148–165. [CrossRef]
11. Boroomand, B.; Bazazzadeh, S.; Zandi, S.M. On the use of Laplace's equation for pressure and a mesh-free method for 3D simulation of nonlinear sloshing in tanks. *Ocean Eng.* **2016**, *122*, 54–67. [CrossRef]
12. Morteza, G.; Mahmoud, G. Numerical analysis of fully non-linear sloshing waves in an arbitrary shape tank by meshless method. *Eng. Anal. Bound. Elem.* **2022**, *144*, 366–369.
13. Sun, X.; Zhong, Y.; Feng, B.; Liu, C.; Yin, Y. Numerical Computation of Sloshing-Induced Force in Complex Ship Tanks under the Excitation of Ship Rolling Motion Based on the MPS Method. *Appl. Sci.* **2022**, *12*, 5130. [CrossRef]
14. Zhang, K.K. Analysis and Anti-Sway Study of Liquid Sloshing in Liquid Tank Based on SPH Method. Master's Thesis, Harbin Institute of Technology, Harbin, China, 2015.
15. Leng, F. Research on Liquid Sloshing in Aircraft Fuel Tank with SPH Technique. Master's Thesis, Nanjing University of Aeronautics and Astronautics, Nanjing, China, 2009.
16. Liu, F. Dynamic Analysis of Liquid Sloshing and Sloshing Suppression Design for a Tank. Ph.D. Thesis, Nanjing University of Aeronautics and Astronautics, Nanjing, China, 2010.
17. Wu, J.L.; Li, J.T.; Deng, Q.L.; Cheng, Y.; Wang, Y.Q. Numerical Research of Structural Damping on the Performance of the U Type Anti-rolling Tank. *Ship Eng.* **2020**, *42*, 80–83. [CrossRef]
18. Luo, H.B.; Wang, C.B.; Sheng, Q.W. Numerical and Model Test on Eigen Characters of Rolling Motion of Anti-Rolling Tank. *China Offshore Platf.* **2020**, *35*, 1–6.
19. Xue, M.A.; Lin, P. Numerical study of ring baffle effects on reducing violent liquid sloshing. *Comput. Fluids* **2011**, *52*, 116–129. [CrossRef]

20. Dasgupta, A. *Effect of Tank Cross-Section and Longitudinal Baffles on Transient Liquid Slosh in Partly-Filled Road Tankers*; Concordia University: Montreal, QC, Canada, 2011.
21. Budiansky, B. Sloshing of liquids in circular canals and spherical tanks. *J. Aerosp. Sci.* **2012**, *27*, 161–173. [CrossRef]
22. Nema, P. Computational Study of Sloshing Behavior in 3-D Rectangular Tank with and Without Baffle under Seismic Excitation. Ph.D. Thesis, National Institute of Technology, Rourkela, India, 2014.
23. Suyal, R. CFD Analysis of Fuel Sloshing in a Cylindrical Tank with and Without Baffles Under Linear Acceleration. Ph.D. Thesis, Concordia University, Montreal, QC, USA, 2016.
24. Guan, Y.; Yang, C.; Chen, P.; Zhou, L. Numerical investigation on the effect of baffles on liquid sloshing in 3D rectangular tanks based on nonlinear boundary element method. *Int. J. Nav. Archit. Ocean Eng.* **2020**, *12*, 399–413. [CrossRef]
25. Liang, L.H.; Wang, J.M.; Song, J.G.; Li, Y. Design and simulation investigation of variable period passive anti-rolling tank. *Chin. J. Ship Res.* **2021**, *16*, 147–154. [CrossRef]
26. Zhao, M.H.; Wang, T. Influence of the Number of Longitudinal Baffles on Liquid Sloshing of Trapezoidal Tank. *Pract. Technol. Automob.* **2022**, *47*, 120–123. [CrossRef]
27. Liu, D.X.; Wang, X.Y.; Chen, Y.J. Numerical, and Analytical Slosh Dynamics of Water and Liquid Nitrogen in a Spherical Tank. *J. Mar. Sci. Eng.* **2024**, *12*, 558. [CrossRef]
28. Zhang, C.L.; Wang, L.Z.; Xu, M. Study on the Damping Effect and Mechanism of Vertical Slotted Screens Based on the BM-MPS Method. *J. Mar. Sci. Eng.* **2023**, *11*, 1270. [CrossRef]
29. Zhang, Q.; Shui, B.; Zhu, H.H. Study on Sloshing Characteristics in a Liquid Cargo Tank under Combination Excitation. *J. Mar. Sci. Eng.* **2022**, *10*, 1100. [CrossRef]
30. Mohapatra, S.C.; Islam, H.; Soares, C.G. Boussinesq Model and CFD Simulations of Non-Linear Wave Diffraction by a Floating Vertical Cylinder. *J. Mar. Sci. Eng.* **2022**, *8*, 575. [CrossRef]
31. Versteeg, H.K.; Malalasekera, W. *An Introduction to Computational Fluid Dynamics: The Finite Volume Method*, 1st ed.; Pearson Education Limited: Chennai, India, 1995.
32. Pan, L.J.; Wu, Z.J.; Zhang, J.F.; Zhang, B.M.; Du, S.Y. Modal analysis of liquid filled in tank applying Rayleigh-Ritz method. *J. Harbin Eng. Univ.* **2007**, *1*, 31–34.
33. Dodge, F.T. *The New "Dynamic Behavior of Liquids in Moving Containers"*; Southwest Research Institute: San Antonio, TX, USA, 2000.

Journal of
*Marine Science and Engineering*

MDPI

*Article*

# Vibration Characteristic Analysis of Sandwich Composite Plate Reinforced by Functionally Graded Carbon Nanotube-Reinforced Composite on Winkler/Pasternak Foundation

**Mengzhen Li [1,2], Xiaolong Liu [3], Mohammad Yazdi [4] and Wei Chen [1,2,*]**

1 Hubei Defense Science and Technology Key Laboratory of Ship Explosion Damage and Protection, Wuhan University of Technology, Wuhan 430063, China
2 School of Naval Architecture, Ocean and Energy Power Engineering, Wuhan University of Technology, Wuhan 430063, China
3 School of Transportation and Logistics Engineering, Wuhan University of Technology, Wuhan 430063, China; xxlong_liu@163.com
4 School of Engineering, Faculty of Science and Engineering, Macquarie University, Sydney, NSW 2109, Australia
* Correspondence: whutcw01@126.com

**Abstract:** This paper presents numerical investigations into the free vibration properties of a sandwich composite plate with two fiber-reinforced plastic (FRP) face sheets and a functionally graded carbon nanotube-reinforced composite (FG-CNTRC) core made of functionally graded carbon nanotube-reinforced composite resting on Winkler/Pasternak elastic foundation. The material properties of the FG-CNTRC core are gradient change along the thickness direction with four distinct carbon nanotubes reinforcement distribution patterns. The Hamilton energy concept is used to develop the equations of motion, which are based on the high-order shear deformation theory (HSDT). The Navier method is then used to obtain the free vibration solutions. By contrasting the acquired results with those using finite elements and with the previous literature, the accuracy of the present approach is confirmed. Moreover, the effects of the modulus of elasticity, the carbon nanotube (CNT) volume fractions, the CNT distribution patterns, the gradient index $p$, the geometric parameters and the dimensionless natural frequencies' elastic basis characteristics are examined. The results show that the FG-CNTRC sandwich composite plate has higher dimensionless frequencies than the functionally graded material (FGM) plate or sandwich plate. And the volume fraction of carbon nanotubes and other geometric factors significantly affect the dimensionless frequency of the sandwich composite plate.

**Keywords:** sandwich composite plates; carbon nanotube-reinforced composite (FG-CNTRC); free vibration; dimensionless fundamental frequency; higher-order shear theory

## 1. Introduction

Sandwich composite plates are widely used in many industrial applications such as the aerospace manufacturing, marine, and machinery fields. In the engineering environment, these structural elements are frequently exposed to erratic loads like ocean waves, wind, and noise. Prolonged vibration can lead to fatigue failure, which can jeopardize structural safety. Thus, from a practical design perspective, it is essential to understand their vibration properties [1–10]. A spring-supported structure is formed by a sequence of independent linear elastic springs in the Winkler–Pasternak model. When examining how structures like buildings, bridges, and machine foundations behave, the Winkler–Pasternak model is especially helpful. It makes it possible for engineers to thoroughly examine a structure's performance and safety in actual soil situations. For instance, in marine structures, the bottom bases of subsea structures (such as the subsea foundations in Figure 1) can be represented as plate structures resting on a Winkler–Pasternak foundation, which can take into account the Winkler spring stiffness of the seabed in the vertical direction and the Pasternak model for resistance to horizontal shear deformation [11]. The Winkler–Pasternak foundation can more accurately evaluate the dynamic characteristics of marine subsea plate structures with soil–structure interactions.

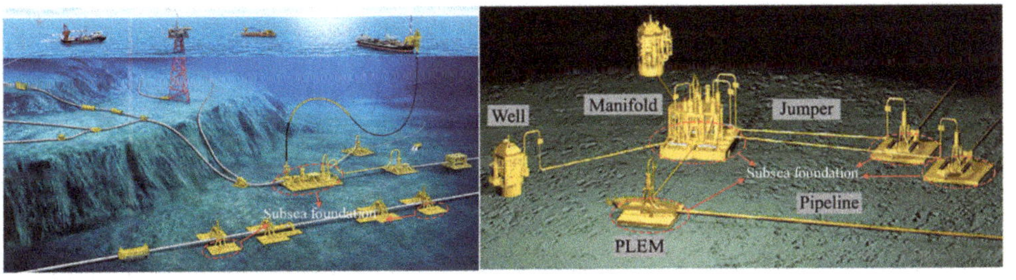


**Figure 1.** Plate-shaped subsea foundation with seabed resistance and soil–structure interactions [12].

Many researchers have examined the free vibration of sandwich plates using the Winkler–Pasternak (WP) model, based on the shear deformation theories [13–18]. Arani and Shariyat [19] assessed the vibrational responses of sandwich panels on WP foundation. Using first-order shear deformation theory (FSDT), Kumar and Harsha [20] studied the bending and diversion of FG panels on elastic foundations. Three-order shear deformation theory (TSDT) was used by Hadji et al. [21] to show the vibrational responses of a FG sandwich plate on an elastic base with an emphasis on geometrical parameters and material distribution. The nonlinear vibration of a sandwich FG plates on a resilient foundations, considering various material parameters was investigated by Li et al. [22]. Shahrany et al. [23] utilized Navier's theoretical solution to study the dynamic response of a sandwich beam on an elastic basis. Selim and Liu [24] used a numerical technique based on HSDT for a parametric study of FG-GPLs on robust foundations. Zaitoun et al. [25] used HSDT to create a modeling strategy and obtain a precise answer for the buckling response of a FG sandwich supported on an elastic foundation. An analysis approach for examining the properties of a sandwich plate's free vibration that is supported by a Pasternak foundation was introduced by Zenkour and El-Shahrany [26]. A novel method for examining the vibration characteristics of sandwich FGM plates resting on an elastic foundation in a heated environment was presented by Singh and Harsha [27]. Shahsavari et al. [28] used an elastic-foundation methodology to conduct multiple parametric investigations for the dynamic responses of FG imperfect plates. In addition, many researchers have also studied numerical methods, like the finite element method. Paik and Seo [29] used the nonlinear finite element method to study the application of ultimate strength to steel plate structures under combined forces. To forecast the steel model's ultimate strength, Putranto et al. [30] carried out a ESL finite element analysis.

Fiber-reinforced plastics (FRPs) are popular and incredibly high-performing composite materials that combine the flexibility and malleability of a matrix resin with the high strength of fibers and are a hot research topic today. They are used extensively in many different industries due to their exceptional mechanical qualities, low weight, and resistance to corrosion. FRPs have many applications, ranging from sports equipment and aerospace to transportation and construction, and their use is growing as technology advances [31].

In recent years, FG-CNTRC has drawn more attention because it eliminates interfacial tensions by gradually changing its material composition as thickness increases [32–37]. Due to their exceptional temperature resistance, lightweight design, and extreme durability, FGM sandwich constructions can withstand high temperatures without losing structural strength or stiffness while also weighing less overall. FGMs are perfect for the nuclear reactor, mechanical engineering, aerospace, and marine industries because of these qualities. This soft, low-density, high-aspect ratio material has developed into a high-quality reinforcing phase in recent years due to the use of carbon nanotubes (CNTs) [38–43]. According to previous research, adding carbon nanotubes greatly enhances the characteristics of metals, ceramics, and polymers. In the research conducted by Mohebpour et al. [44], a numerical approach was presented to study dynamic responses of FG-CNTRC pipes. The vibrational responses of FG-CNTRC, indicating the impacts of CNT volume fractions and distribution patterns on vibration behavior were investigated by Tayeb et al. [45]. Cho [46] presented a numerical analysis based on element method to research the large deflection static problem of FG-CNTRC plate on elastic basis. Duong et al. [47] conducted a thorough investigation of the stress centralization phenomena and the static analysis of FG-CNTRC cylindrical shells. Duc and Minh [48] investigated the free vibration behavior of cracked FG-CNTRC plates using shear deformation theory and finite element method. Taheri and Memarzadeh [49] studied the dynamic behavior of cracked CNTRC plates under load and investigated the effect of various parameters on the vibrational behavior. Zeighami and Jafari [50] used a numerical solution to analyze thermal stresses of FG-CNTRC plate. Cong et al.'s study [51] examined the behavior of the laminated double-curved thin shell made of FG-CNTRC under different geometrical conditions. The vibration properties of a sandwich comprising FG-CNTRC panels and a metal core were examined by Watts et al. [52]. Duc and Minh [48] used TSDT to study the dynamic characteristics of FG-CNTRC plates with cracks. A high-order discretization was presented by Lin et al. [53] to study the FG-CNTRC beam's random vibration behavior. The aforementioned research has demonstrated that carbon nanotubes can greatly enhance a material's mechanical qualities. However, the application of CNTs in FRP-FGM-FRP sandwich plates has never been developed, and its kinetic properties are not yet clear.

Therefore, the novelty and importance of this work are found in the functional core layer that is created when CNTs are added to FGMs made of metals and ceramics to create the FG-CNTRC plate. This further enhances the sandwich composite plate's dimensionless fundamental frequency. Furthermore, it is believed that the Winkler/Pasternak foundation is the location of this plate. Hamilton's principle and HSDT were used to develop differential equations, and Navier's method was used to obtain the dynamic results. Moreover, the impact of several aspects on the non-dimensional frequency of the sandwich plate are examined, including geometric parameters, gradient index, carbon nanotube volume percentage, carbon nanotube dispersal type, and elastic foundation parameters.

## 2. Theoretical Formulation

### 2.1. Description of the Model

In this paper, a rectangular sandwich composite plate in a Cartesian coordinate system (*x*-*y*-*z*) with two fiber-reinforced plastics (FRP) face sheets and a novel hybrid composite core is considered. The hybrid core consists of carbon nanotube-reinforced composites (CNTRCs) and metal–ceramic functionally graded (FG) matrix. Winkler and Pasternak foundations support the sandwich plate, which has dimensions of $a$, $b$, and $h$. Furthermore, the core and face sheets have thicknesses of $h_c$ and $h_f$, respectively. The schematic diagram of the structure is shown in Figure 2.

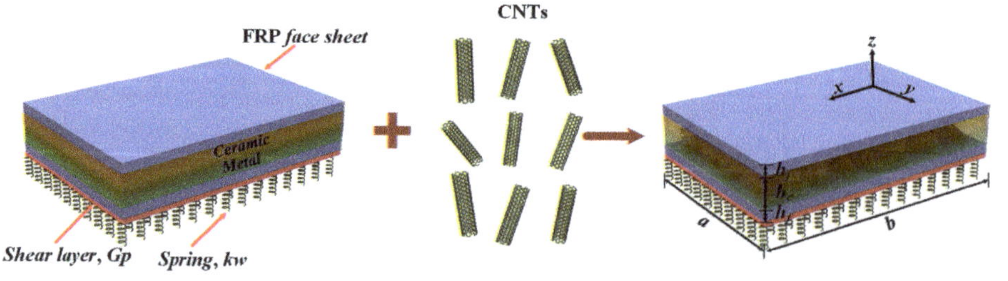


**Figure 2.** Configuration of sandwich plates.

### 2.2. Material Properties of CNT-Reinforced Plates

The material composition of the FG plate progressively changes from all-ceramic on the upper surface to all-metal on the bottom surface, with material characteristics varying in accordance with power law regulations [54]. The hybrid composite core's matrix is made of metal–ceramic functionally graded materials. For instance, the mixing rule determines the power law distribution of the effective modulus of elasticity and density, which varies continuously along the thickness direction. The core matrix's material properties are shown in Equation (1a,b). For the FGM core, Poisson's coefficient is assumed to be constant (i.e., it is independent of the vertical coordinate $z$). Moreover, Young's modulus $E$ and mass density $p$ are independent of temperature.

$$E_m(z) = (E_t - E_b)(\frac{z}{h_c} + \frac{1}{2})^p + E_b \tag{1a}$$

$$\rho_m(z) = (\rho_t - \rho_b)(\frac{z}{h_c} + \frac{1}{2})^p + \rho_b \tag{1b}$$

where $E_m$ and $\rho_m$ denote the elasticity modulus and density of the interlayer matrix; $E_t$ and $E_b$ denote the elasticity modulus of the ceramic and metal; $\rho_t$ and $\rho_b$ denote the density of the ceramic and metal, respectively; and $p$ is the gradient index.

In this research, four types of CNT distributions (UD, FG-V, FG-O, and FG-X) along the thickness direction of the intermediate layer are investigated, as seen in Figure 3. The volume fractions of CNTs with different distribution types are shown in Equation (2a–d):

$$\text{UD}: V_{CNT}(z) = V_{CNT}^* \tag{2a}$$

$$\text{FG-V}: V_{CNT}(z) = (1 + \frac{2z}{h_c})V_{CNT}^* \tag{2b}$$

$$\text{FG-O}: V_{CNT}(z) = 2(1 - \frac{2|z|}{h_c})V_{CNT}^* \tag{2c}$$

$$\text{FG-X}: V_{CNT}(z) = 2(\frac{2|z|}{h_c})V_{CNT}^* \tag{2d}$$

where

$$V_{CNT}^*(z) = \frac{\omega_{CNT}}{\omega_{CNT} + \frac{\rho_{CNT}}{\rho_m} - \frac{\rho_{CNT}}{\rho_m}\omega_{CNT}} \tag{3}$$

In which $\omega_{CNT}$ denotes the mass fraction of carbon nanotubes; $\rho_{CNT}$ and $\rho_m$ are the densities of the carbon nanotubes and matrix, respectively; and $z$ is the interlayer thickness coordinate.

The effective elastic modulus, shear modulus, Poisson's ratio, and mass density of the hybrid composite core in this work can be obtained from Equation (4a–f) by incorporating the efficiency parameters of CNTs:

$$E_{11}(z) = \eta_1 V_{CNT}(z)E_{11}^{CNT} + V_m E^m \tag{4a}$$

$$\frac{\eta_2}{E_{22}(z)} = \frac{V_{CNT}(z)}{E_{22}^{CNT}} + \frac{V_m(z)}{E^m} \tag{4b}$$

$$\frac{\eta_3}{G_{12}(z)} = \frac{V_{CNT}(z)}{G_{12}^{CNT}} + \frac{V_m(z)}{G^m} \tag{4c}$$

$$\mu_{12}(z) = V_{CNT}(z)\mu_{12}^{CNT} + V_m(z)\mu^m \tag{4d}$$

$$\mu_{21}(z) = \frac{\mu_{12}(z)}{E_{11}(z)}E_{22}(z) \tag{4e}$$

$$\rho(z) = V_{CNT}(z)\rho^{CNT} + V_m(z)\rho^m \tag{4f}$$

where $E_{11}^{CNT}$, $E_{22}^{CNT}$, $G_{12}^{CNT}$, $\mu_{12}^{CNT}$, and $\rho^{CNT}$ denote the longitudinal and transverse elastic modulus, shear modulus, Poisson's ratio, and density of carbon nanotubes, respectively. $E^m$, $G^m$, $\mu^m$, and $\rho^m$ denote the corresponding properties of the isotropic FGM matrix, respectively. $\eta_1$, $\eta_2$, and $\eta_3$ are the efficiency parameters considering the three directions of CNTs, and the values of the CNTs efficiency parameters used according to Shen [55] are shown in Table 1. $V_{CNT}$ and $V_m$ denote the volume fractions of the CNTs and FGM matrix materials, and we obtain $V_{CNT}$ + $V_m$ =1.

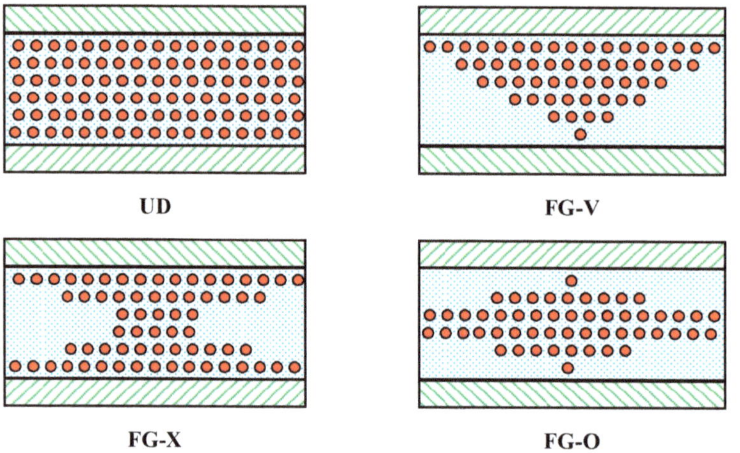


**Figure 3.** Configuration of various FG-CNTRC rectangular plates (The red circles are carbon nanotubes).

**Table 1.** Three different nanotube volume fractions' values for the efficiency parameters [55].

| $V_{CNT}$ | $\eta_1$ | $\eta_2$ | $\eta_3$ |
|---|---|---|---|
| 0.12 | 0.137 | 1.022 | 0.715 |
| 0.17 | 0.142 | 1.626 | 1.138 |
| 0.28 | 0.141 | 1.585 | 1.109 |

*2.3. HSDT-Based Models*

The HSDT-based displacements of the novel sandwich composite plate along $x$, $y$, and $z$ can be given as

$$\left\{\begin{array}{c} u(x,y,z,t) \\ v(x,y,z,t) \\ w(x,y,z,t) \end{array}\right\} = \left\{\begin{array}{c} u_0(x,y,t) \\ v_0(x,y,t) \\ w_0(x,y,t) \end{array}\right\} - \left\{\begin{array}{c} z\frac{\partial w_0}{\partial x} \\ z\frac{\partial w_0}{\partial y} \\ 0 \end{array}\right\} + \left\{\begin{array}{c} f(z)\theta_x \\ f(z)\theta_y \\ 0 \end{array}\right\} \tag{5}$$

where the midplane surface is indicated by the subscript 0, and the displacements in the $x$, $y$, and $z$ directions are indicated by $u$, $v$, and $w$. The transverse shear shape function is denoted by $f(z)$. Rotations of the midplane normal along the $x$ and $y$ axes are represented by the numbers $\theta_x$ and $\theta_y$. Equation (6) can be used to express the strain at a location in the plate when taking into account transverse and in-plane shear strains.

$$\varepsilon = \varepsilon^0 + z\varepsilon^b + f(z)\varepsilon^s; \gamma = f\prime(z)\gamma^0 \tag{6}$$

where

$$\varepsilon^0 = \begin{Bmatrix} \varepsilon_x^0 \\ \varepsilon_y^0 \\ \gamma_{xy}^0 \end{Bmatrix} = \begin{Bmatrix} \frac{\partial u_0}{\partial x} \\ \frac{\partial v_0}{\partial y} \\ \frac{\partial v_0}{\partial x} + \frac{\partial u_0}{\partial y} \end{Bmatrix}, \varepsilon^b = \begin{Bmatrix} \varepsilon_x^b \\ \varepsilon_y^b \\ \gamma_{xy}^b \end{Bmatrix} = \begin{Bmatrix} -\frac{\partial^2 w_0}{\partial x^2} \\ -\frac{\partial^2 w_0}{\partial y^2} \\ -2\frac{\partial^2 w_0}{\partial x \partial y} \end{Bmatrix}, \varepsilon^s = \begin{Bmatrix} \varepsilon_x^s \\ \varepsilon_y^s \\ \gamma_{xy}^s \end{Bmatrix} = \begin{Bmatrix} \frac{\partial \theta_x}{\partial x} \\ \frac{\partial \theta_y}{\partial y} \\ \frac{\partial \theta_x}{\partial y} + \frac{\partial \theta_y}{\partial x} \end{Bmatrix},$$
$$\gamma^0 = \begin{Bmatrix} \gamma_{xz}^0 \\ \gamma_{yz}^0 \end{Bmatrix} = \begin{Bmatrix} \theta_x \\ \theta_y \end{Bmatrix} \tag{7}$$

where $\varepsilon_x$ and $\varepsilon_y$ are the normal strains, $\gamma_{xy}$ is the in-plane shear strain, $\gamma_{xz}$ and $\gamma_{yz}$ are the transverse shear strains. In this paper, due to the anisotropic FRP face sheets and the isotropic core, their link between stress and strain can be stated as

$$\begin{Bmatrix} \sigma_x \\ \sigma_y \\ \tau_{xy} \\ \tau_{yz} \\ \tau_{xz} \end{Bmatrix}^{(k)} = \begin{bmatrix} \overline{Q}_{11}^{(k)} & \overline{Q}_{12}^{(k)} & \overline{Q}_{16}^{(k)} & 0 & 0 \\ \overline{Q}_{12}^{(k)} & \overline{Q}_{22}^{(k)} & \overline{Q}_{26}^{(k)} & 0 & 0 \\ \overline{Q}_{16}^{(k)} & \overline{Q}_{26}^{(k)} & \overline{Q}_{66}^{(k)} & 0 & 0 \\ 0 & 0 & 0 & \overline{Q}_{44}^{(k)} & \overline{Q}_{45}^{(k)} \\ 0 & 0 & 0 & \overline{Q}_{45}^{(k)} & \overline{Q}_{55}^{(k)} \end{bmatrix} \begin{Bmatrix} \varepsilon_x \\ \varepsilon_y \\ \gamma_{xy} \\ \gamma_{yz} \\ \gamma_{xz} \end{Bmatrix}^{(k)} \tag{8a}$$

$$\begin{Bmatrix} \sigma_x \\ \sigma_y \\ \tau_{yz} \\ \tau_{xz} \\ \tau_{xy} \end{Bmatrix} = \begin{bmatrix} Q_{11} & Q_{12} & 0 & 0 & 0 \\ Q_{12} & Q_{22} & 0 & 0 & 0 \\ 0 & 0 & Q_{44} & 0 & 0 \\ 0 & 0 & 0 & Q_{55} & 0 \\ 0 & 0 & 0 & 0 & Q_{66} \end{bmatrix} \begin{Bmatrix} \varepsilon_x \\ \varepsilon_y \\ \gamma_{yz} \\ \gamma_{xz} \\ \gamma_{xy} \end{Bmatrix} \tag{8b}$$

where $\sigma_x$ and $\sigma_y$ are the in-plane normal stresses, $\tau_{xy}$ is the in-plane shear stress, and $\tau_{xz}$ and $\tau_{yz}$ are the transverse shear stresses. $k$ denotes the $k$th layer of the FRP plate. The stiffness coefficients $Q_{ij}$ for the FRP face sheets are expressed as

$$Q_{11} = Q_{22} = \frac{E(z)}{1-\mu^2}, Q_{12} = \frac{\mu E(z)}{1-\mu^2}, Q_{44} = Q_{55} = Q_{66} = \frac{E(z)}{2(1+\mu)} \tag{9}$$

and

$$\begin{aligned} \overline{Q}_{11}^{(k)} &= Q_{11}\cos^4\alpha_k + 2(Q_{12}+2Q_{66})\sin^2\alpha_k\cos^2\alpha_k + Q_{22}\sin^4\alpha_k \\ \overline{Q}_{12}^{(k)} &= Q_{12}(\sin^4\alpha_k+\cos^4\alpha_k) + (Q_{11}+Q_{22}-4Q_{66})\sin^2\alpha_k\cos^2\alpha_k \\ \overline{Q}_{22}^{(k)} &= Q_{11}\cos^4\alpha_k + 2(Q_{12}+2Q_{66})\sin^2\alpha_k\cos^2\alpha_k + Q_{11}\sin^4\alpha_k \\ \overline{Q}_{16}^{(k)} &= (Q_{11}-Q_{12}-2Q_{66})\sin\alpha_k\cos^3\alpha_k + (Q_{12}-Q_{22}+2Q_{66})\sin^3\alpha_k\cos\alpha_k \\ \overline{Q}_{26}^{(k)} &= (Q_{11}-Q_{22}+2Q_{66})\sin\alpha_k\cos^3\alpha_k + (Q_{11}-Q_{12}-2Q_{66})\sin^3\alpha_k\cos\alpha_k \\ \overline{Q}_{66}^{(k)} &= Q_{66}(\sin^4\alpha_k+\cos^4\alpha_k) + (Q_{11}+Q_{22}-2Q_{12}-2Q_{66})\sin^2\alpha_k\cos^2\alpha_k \\ \overline{Q}_{44}^{(k)} &= Q_{44}\cos^2\alpha_k + Q_{55}\sin^2\alpha_k \\ \overline{Q}_{45}^{(k)} &= (Q_{55}-Q_{44})\sin\alpha_k\cos\alpha_k \\ \overline{Q}_{55}^{(k)} &= Q_{55}\cos^2\alpha_k + Q_{44}\sin^2\alpha_k \end{aligned} \tag{10}$$

where $\alpha_k$ denotes the angle of fiber formation in the $k$th layer. The stiffness coefficients $Q_{ij}$ for the hybrid composite core are expressed as

$$Q_{11} = \frac{E_{11}(z)}{1-\mu_{12}\mu_{21}}, Q_{12} = \frac{\mu_{21}E_{11}(z)}{1-\mu_{12}\mu_{21}}, Q_{22} = \frac{E_{22}(z)}{1-\mu_{12}\mu_{21}}, Q_{44} = G_{23}, Q_{55} = G_{13}, Q_{66} = G_{12} \tag{11}$$

The forces and moments of the sandwich plate can be written in Equation (12).

$$\left\{\begin{array}{c} N_x \\ M_x \\ S_x \\ N_y \\ M_y \\ S_y \\ N_{xy} \\ M_{xy} \\ S_{xy} \end{array}\right\} = \begin{bmatrix} A_{11} & A_{12} & A^k_{16} & B_{11} & B_{12} & B^k_{16} & C_{11} & C_{12} & C^k_{16} \\ B_{11} & B_{12} & B^k_{16} & D_{11} & D_{12} & D^k_{16} & E_{11} & E_{12} & E^k_{16} \\ C_{11} & C_{12} & C^k_{16} & E_{11} & E_{12} & E^k_{16} & F_{11} & F_{12} & F^k_{16} \\ A_{12} & A_{22} & A^k_{26} & B_{12} & B_{22} & B^k_{26} & C_{12} & C_{22} & C^k_{26} \\ B_{12} & B_{22} & B^k_{26} & D_{12} & D_{22} & D^k_{26} & E_{12} & E_{22} & E^k_{26} \\ C_{12} & C_{22} & C^k_{26} & E_{12} & E_{22} & E^k_{26} & F_{12} & F_{22} & F^k_{26} \\ A^k_{16} & A^k_{26} & A_{66} & B^k_{16} & B^k_{26} & B_{66} & C^k_{16} & C^k_{26} & C_{66} \\ B^k_{16} & B^k_{26} & B_{66} & D^k_{16} & D^k_{26} & D_{66} & E^k_{16} & E^k_{26} & E_{66} \\ C^k_{16} & C^k_{26} & C_{66} & E^k_{16} & E^k_{26} & E_{66} & F^k_{16} & F^k_{26} & F_{66} \end{bmatrix} \left\{\begin{array}{c} \varepsilon^0_x \\ \varepsilon^0_y \\ \gamma^0_{xy} \\ \varepsilon^b_x \\ \varepsilon^b_y \\ \gamma^b_{xy} \\ \varepsilon^s_x \\ \varepsilon^s_y \\ \gamma^s_{xy} \end{array}\right\}, \left\{\begin{array}{c} N_{xz} \\ N_{yz} \end{array}\right\} = \begin{bmatrix} G_{55} & G^k_{45} \\ G^k_{45} & G_{44} \end{bmatrix} \left\{\begin{array}{c} \gamma^0_{xz} \\ \gamma^0_{yz} \end{array}\right\} \quad (12)$$

where $P^k_{ij}$ ($P = A, B, C, D, E, F$ and G) = 0 when $k$ = CNT, and $P \neq 0$ when $p$ = FRP. These coefficients can be specified as

$$\begin{aligned} (A_{ij}, B_{ij}, C_{ij}, D_{ij}, E_{ij}, F_{ij}, G_{ij}) = \eta_{ij} = \eta^t_{ij} + \eta^c_{ij} + \eta^b_{ij} = \int_{-\frac{h}{2}}^{-\frac{h_c}{2}} \overline{Q}^{(k)}_{ij} \{1, z, f(z), z^2, zf(z), f^2(z), f\prime(z)\} dz \\ + \int_{-\frac{h_c}{2}}^{\frac{h_c}{2}} Q_{ij}\{1, z, f(z), z^2, zf(z), f^2(z), f\prime(z)\} dz + \int_{\frac{h_c}{2}}^{\frac{h}{2}} \overline{Q}^{(k)}_{ij} \{1, z, f(z), z^2, zf(z), f^2(z), f\prime(z)\} dz \end{aligned} \quad (13)$$

where $\eta$ = A, B, C, D, E, F, and G; $i$, $j$ = 1, 2, 4, 5 and 6; and the superscript "$t$", "$c$", and "$b$" represent the sandwich plate's top, middle, and bottom layers, respectively. The governing equations of motion for the sandwich plate can be determined by using Hamilton's principle, which is given by

$$\int_0^t (\delta U + \delta V_e - \delta T) dt = 0 \quad (14)$$

where the variations in strain energy are denoted by $\delta U$, the elastic basis's potential energy is represented by $\delta V_e$, and the kinetic energy is represented by $\delta T$. The plate's strain energy can be described as

$$\begin{aligned} \delta U &= \int_A \int_{-h/2}^{h/2} (\sigma_x \delta\varepsilon_x + \sigma_y \delta\varepsilon_y + \sigma_{xy}\delta\gamma_{xy} + \sigma_{xz}\delta\gamma_{xz} + \sigma_{yz}\delta\gamma_{yz}) dA dz \\ &= \int_A [N_x \frac{\partial \delta u_0}{\partial x} - M_x \frac{\partial^2 \delta w_0}{\partial x^2} + S_x \frac{\partial \delta\theta_x}{\partial x} + N_y \frac{\partial \delta v_0}{\partial y} - M_y \frac{\partial^2 \delta w_0}{\partial y^2} + S_y \frac{\partial \delta\theta_y}{\partial y} + N_{xy}(\frac{\partial \delta u_0}{\partial y} + \frac{\partial \delta v_0}{\partial x}) \\ &\quad -2M_{xy} \frac{\partial^2 \delta w_0}{\partial x \partial y} + S_{xy}(\frac{\partial \delta\theta_x}{\partial y} + \frac{\partial \delta\theta_y}{\partial x}) + N_{xz}\delta\theta_x + N_{yz}\delta\theta_y] dA \end{aligned} \quad (15)$$

For the variation in elastic potential energy, it can be expressed as

$$\delta V_e = \int_A f_e \, \delta w_0 dA \quad (16)$$

where $f_e$ is specifically expressed in the form [56]

$$\begin{aligned} f_e(Winkler) &= q_{Winkler} = K_w w_0, \\ f_e(Pasternak) &= q_{Pasternak} = K_w w_0 - G_p \nabla^2 w_0 \end{aligned} \quad (17)$$

where $\nabla^2 = \partial^2/\partial x^2 + \partial^2/\partial y^2$ is the Laplace operator in a rectangular Cartesian coordinate system [57]. The kinetic energy of the plate is specified as

$$
\begin{aligned}
&T = \int_A \int_{-h/2}^{h/2} (\dot{u}\delta\dot{u} + \dot{v}\delta\dot{v} + \dot{w}\delta\dot{w})\rho(z)dzdA \\
&\int_A \Big\{ I_0(\dot{u}_0\delta\dot{u}_0 + \dot{v}_0\delta\dot{v}_0 + \dot{w}_0\delta\dot{w}_0) - I_1(\dot{u}_0\frac{\partial\delta\dot{w}_0}{\partial x} + \frac{\partial\dot{w}_0}{\partial x}\delta\dot{u}_0 + \dot{v}_0\frac{\partial\delta\dot{w}_0}{\partial x} + \frac{\partial\dot{w}_0}{\partial x}\delta\dot{v}_0) \\
&+I_2(\frac{\partial\dot{w}_0}{\partial x}\frac{\partial\delta\dot{w}_0}{\partial x} + \frac{\partial\dot{w}_0}{\partial y}\frac{\partial\delta\dot{w}_0}{\partial y}) + I_3(\dot{u}_0\delta\dot{\theta}_x + \dot{\theta}_x\delta\dot{u}_0 + \dot{v}_0\delta\dot{\theta}_y + \dot{\theta}_y\delta\dot{v}_0) + I_4(\dot{\theta}_x\delta\dot{\theta}_x + \dot{\theta}_y\delta\dot{\theta}_y) \\
&-I_5(\frac{\partial\dot{w}_0}{\partial x}\delta\dot{\theta}_x + \dot{\theta}_x\frac{\partial\delta\dot{w}_0}{\partial x} + \frac{\partial\dot{w}_0}{\partial y}\delta\dot{\theta}_y + \dot{\theta}_y\frac{\partial\delta\dot{w}_0}{\partial y})\Big\}dA
\end{aligned}
\tag{18}
$$

where the differentiation with regard to the time variable is indicated by the superscripted dot, and $\rho(z)$ denotes the density along the thickness direction $z$. $I_0, I_1, I_2, I_3, I_4, I_5$ are the inertia terms, which are expressed as

$$
\begin{aligned}
&(I_0, I_1, I_2, I_3, I_4, I_5,) = \int_{-h/2}^{-h_c/2} [1, z, z^2, f(z), f(z)z, f^2(z)]\rho(z)dz \\
&+\int_{-h_c/2}^{h_c/2} [1, z, z^2, f(z), f(z)z, f^2(z)]\rho(z)dz + \int_{h_c/2}^{h/2} [1, z, z^2, f(z), f(z)z, f^2(z)]\rho(z)dz
\end{aligned}
\tag{19}
$$

Substituting Equations (15), (16), and (18) into Equation (14), the governing equations for the plate can be derived using the Hamiltonian energy principle and collected $\delta u_0$,$\delta v_0$, $\delta w_0$, $\delta\theta_x$ and $\delta\theta_y$, respectively denoted as

$$
\delta u_0 : \frac{\partial N_x}{\partial x} + \frac{\partial N_{xy}}{\partial y} = I_0\ddot{u}_0 - I_1\frac{\partial\ddot{w}}{\partial x} - I_3\frac{\partial\theta}{\partial x}
\tag{20a}
$$

$$
\delta v_0 : \frac{\partial N_y}{\partial y} + \frac{\partial N_{xy}}{\partial x} = I_0\ddot{v}_0 - I_1\frac{\partial\ddot{w}}{\partial y} - I_3\frac{\partial\ddot{\theta}}{\partial y}
\tag{20b}
$$

$$
\delta w_0 : \frac{\partial^2 M_x}{\partial x^2} + \frac{\partial^2 M_y}{\partial y^2} + 2\frac{\partial^2 M_{xy}}{\partial x\partial y} - f_e = I_0\ddot{w}_0 + I_1(\frac{\partial\ddot{u}}{\partial x} + \frac{\partial\ddot{v}}{\partial y}) - I_4(\frac{\partial^2\ddot{\theta}_x}{\partial x^2} + \frac{\partial^2\ddot{\theta}_y}{\partial y^2}) + I_0\ddot{w}_0
\tag{20c}
$$

$$
\delta\theta_x : \frac{\partial s_x}{\partial x} + \frac{\partial s_{xy}}{\partial y} - N_{xz} = I_2\ddot{u}_0 - I_4\frac{\partial\ddot{w}_0}{\partial x} + I_5\ddot{\theta}_x
\tag{20d}
$$

$$
\delta\theta_y : \frac{\partial s_y}{\partial y} + \frac{\partial s_{xy}}{\partial x} - N_{yz} = I_2\ddot{v}_0 - I_4\frac{\partial\ddot{w}_0}{\partial y} + I_5\ddot{\theta}_y
\tag{20e}
$$

Equation (12) is substituted into Equation (20a)–(20e), and by replacing the coefficients in the equation, expansions of the equation $\delta u_0$, $\delta v_0$, $\delta w_0$, $\delta\theta_x$, and $\delta\theta_y$ can be derived, and these expansions are listed in Appendix A.

### 2.4. Solutions for Simply Supported Sandwich Plates

The displacement variables for the sandwich plate to satisfy the boundary conditions are given as follows:

$$
\begin{aligned}
&u_0 = U_{ij}\cos(\lambda x)\sin(\mu y)e^{iwt}, v_0 = V_{ij}\sin(\lambda x)\cos(\mu y)e^{iwt}, w_0 = W_{ij}\sin(\lambda x)\sin(\mu y)e^{iwt} \\
&\theta_x = X_{ij}\cos(\lambda x)\sin(\mu y)e^{iwt}, \theta_y = Y_{ij}\sin(\lambda x)\cos(\mu y)e^{iwt}
\end{aligned}
\tag{21}
$$

where $U_{ij}$, $V_{ij}$, $W_{ij}$, $X_{ij}$, and $Y_{ij}$ are the five unknown parameters, $w$ represents the natural frequency. The variables $\lambda = i\pi/a$, $\mu = j\pi/b$, where $i$ and $j$ are truncated numbers. For sinusoidally distributed loads, the following equation is obtained by substituting Equation (21) into the expansions of the equation, i.e., $\delta u_0$, $\delta v_0$, $\delta w_0$, $\delta\theta_x$, and $\delta\theta_y$:

$$
\left(\begin{bmatrix} P_{11} & P_{12} & P_{13} & P_{14} & P_{15} \\ P_{21} & P_{22} & P_{23} & P_{24} & P_{25} \\ P_{31} & P_{32} & P_{33} & P_{34} & P_{35} \\ P_{41} & P_{42} & P_{43} & P_{44} & P_{45} \\ P_{51} & P_{52} & P_{53} & P_{54} & P_{55} \end{bmatrix} - w^2 \begin{bmatrix} Q_{11} & Q_{12} & Q_{13} & Q_{14} & Q_{15} \\ Q_{21} & Q_{22} & Q_{23} & Q_{24} & Q_{25} \\ Q_{31} & Q_{32} & Q_{33} & Q_{34} & Q_{35} \\ Q_{41} & Q_{42} & Q_{43} & Q_{44} & Q_{45} \\ Q_{51} & Q_{52} & Q_{53} & Q_{54} & Q_{55} \end{bmatrix}\right) \begin{Bmatrix} U_{ij} \\ V_{ij} \\ W_{ij} \\ X_{ij} \\ Y_{ij} \end{Bmatrix} = \begin{Bmatrix} 0 \\ 0 \\ 0 \\ 0 \\ 0 \end{Bmatrix}
\tag{22}
$$

where

$$P_{11} = -A_{11}\lambda^2 - 2A_{16}\lambda\mu - A_{66}\mu^2,\ P_{12} = -A_{12}\lambda\mu - A_{16}\lambda^2 - A_{26}\mu^2 - A_{66}\lambda\mu,\ P_{13} = B_{11}\lambda^3 + B_{12}\lambda\mu^2 + 3B_{16}\lambda^2\mu + B_{26}\mu^3 + 2B_{66}\lambda\mu^2,\ P_{14} = -C_{11}\lambda^2 - 2C_{16}\lambda\mu - C_{66}\mu^2,\ P_{15} = -C_{12}\lambda\mu - C_{16}\lambda^2 - C_{26}\mu^2 - C_{66}\lambda\mu, P_{22} = -A_{22}\mu^2 - 2A_{26}\lambda\mu - A_{66}\lambda^2, P_{23} = B_{12}\lambda^2\mu + 3B_{26}\lambda\mu^2 + B_{16}\lambda^3 + 2B_{66}\lambda^2\mu + B_{22}\mu^3,\ P_{24} = -C_{12}\lambda\mu - C_{26}\mu^2 - C_{16}\lambda^2 - C_{66}\lambda\mu,\ P_{25} = -C_{22}\mu^2 - 2C_{26}\lambda\mu - C_{66}\lambda^2, P_{33} = -D_{11}\lambda^4 - 2D_{12}\lambda^2\mu^2 + 4D_{16}\lambda^3\mu + 4D_{26}\lambda\mu^3 - 4D_{66}\lambda^2\mu^2 \underbrace{-D_{22}\mu^4 - \underbrace{k_w}_{q_{Winker}} - G_p(\lambda^2 + \mu^2)}_{q_{Pasternak}},\ P_{34} = E_{11}\lambda^3 - 3E_{16}\lambda^2\mu - E_{26}\mu^3 + E_{12}\lambda\mu^2 + 2E_{66}\lambda\mu^2, P_{35} = E_{12}\lambda\mu^2 - E_{16}\lambda^3 - 3E_{26}\lambda\mu^2 + E_{22}\mu^3 + 2E_{66}\lambda^2\mu, P_{44} = -F_{11}\lambda^2 - 2F_{16}\lambda\mu - F_{66}\mu^2 - G_{55}, P_{45} = -F_{12}\lambda\mu - F_{16}\lambda^2 - F_{26}\mu^2 - F_{66}\lambda\mu - G_{45}, P_{55} = -G_{44} - F_{22}\mu^2 - F_{66}\lambda^2 - 2F_{26}\lambda\mu \quad (23)$$

in which

$$Q_{11} = -I_0,\ Q_{12} = 0,\ Q_{13} = I_1\lambda,\ Q_{14} = -I_3,\ Q_{15} = 0,\ Q_{22} = -I_0,\ Q_{23} = I_1\mu,\ Q_{24} = 0,\ Q_{25} = -I_3,\ Q_{33} = -I_0 - I_2\lambda^2 - I_2\mu^2,\ Q_{34} = I_4\lambda,\ Q_{35} = I_4\mu,\ Q_{44} = -I_5,\ Q_{45} = 0,\ Q_{55} = -I_5 \quad (24)$$

## 3. Results and Discussion

This first section is to confirm accuracy of the present method before the impact of a number of parameters on the free vibration intrinsic frequency of sandwich plates is conducted, including the modulus of elasticity, gradient index, *kw*, *Gp*, aspect ratio, volume fraction, core thickness ratio ($h_c/h$), and type of distribution of carbon nanotubes. Regarding the transverse shear function $f(z)$ of the displacement field, the shear shape function proposed by Li [15] is used in this study, which is expressed in the form of $f(z) = 8.169z - 7.169h\log[\sec(\frac{z}{h}) + \tan(\frac{z}{h})]$, and its derivative $g(z)$ is denoted as $g(z) = 8.169 - 7.169\sec(\frac{z}{h})$. The geometric parameters used in this analysis are $a = b = 1$ m, $h_f = 0.01$ m, $h_c = 0.1$ m, $kw = Gp = 100$; the fiber layer is [0/90/90/0]; and the CNT type is UD. Table 2 [58], Table 3 [15], and Table 4 [38] show the material properties.

**Table 2.** Material properties of FRP sheet plate.

| Material | $E_{11}^{FRP}$ (GPa) | $E_{22}^{FRP}$ (GPa) | $\mu_{12}^{FRP}$ | $G_{12}^{FRP}$ (GPa) | $G_{13}^{FRP}$ (GPa) | $G_{12}^{FRP}$ (GPa) | $G_{23}^{FRP}$ (GPa) |
|---|---|---|---|---|---|---|---|
| FRP | 40 | 1 | 0.25 | 0.6 | 0.6 | 0.6 | 0.5 |

**Table 3.** Material properties of FGM plate.

| Material | $E_b$ (GPa) | $\rho_b$ (kg/m$^3$) | $E_t$ (GPa) | $\rho_t$ (kg/m$^3$) | $\mu_b$ | $\mu_t$ |
|---|---|---|---|---|---|---|
| FGM | 70 | 2702 | 210 | 2400 | 0.3 | 0.3 |

**Table 4.** Material properties of the CNTs.

| Material | $E_{11}^{cnt}$ (GPa) | $E_{22}^{cnt}$ (GPa) | $G_{12}^{cnt}$ (GPa) | $\mu_{12}^{cnt}$ | $\rho^{cnt}$ (kg/m$^3$) | $V_{CNT}$ |
|---|---|---|---|---|---|---|
| CNT | 5.6466 | 70.8 | 1.9445 | 0.175 | 1150 | 0.12 |

### 3.1. Comparison and Validation

To demonstrate the accuracy and practicality of the proposed approach, this section provides several situations that contrast the obtained results with finite element and the literature results. For convenience, the following intrinsic frequencies are in a dimensionless form: $\widetilde{\omega} = \omega h\sqrt{\rho_b/E_b}$. The formula for calculating the error is $Diff. = \left|(\widetilde{\omega} - \widetilde{\omega}_{\mathrm{Ref}})/\widetilde{\omega}_{\mathrm{Ref}}\right| \times 100$.

Table 5 compares the dimensionless fundamental frequencies of a simply supported FG plate on a Winkler–Pasternak foundation with gradient index (0, 0.5, 1, 2, and 5) and thickness-to-width ratios ($h/a$ = 0.05, 0.1, 0.15, and 0.2) for various elastic foundation parameters. Noting that the following material attributes are employed in this computation:

$E_b$ = 70 GPa, $\rho_b$ = 2702 kg/m$^3$, $E_t$ = 380 GPa, $\rho_t$ = 3800 kg/m$^3$, and $\mu_b = \mu_t = 0.3$. As seen in Table 5, the present results are consistent with the quasi-3D HSDT results reported by Hasani et al. [59], Shahsavari et al. [28], and Li et al. [15]. Furthermore, there is greater agreement between the current numerical results and the quasi-3D results in ref. [15]. The dimensionless intrinsic frequency of simply supported metal–ceramic FGM plates on the Winkler–Pasternak foundation may be accurately calculated using the current algorithm.

**Table 5.** Dimensionless fundamental frequencies $\tilde{\omega} = \omega h\sqrt{\rho_b/E_b}$ of the FG plates.

| *kw* | *Gp* | *h*/*a* | Model | *p* | | | | |
|---|---|---|---|---|---|---|---|---|
| | | | | 0 | 0.5 | 1 | 2 | 5 |
| 0 | 0 | 0.1 | TSDT [59] | 0.1134 | 0.0975 | 0.0891 | 0.0819 | 0.0767 |
| | | | Quasi-3D [28] | 0.1135 | 0.0970 | 0.0882 | 0.0806 | 0.0755 |
| | | | Li et al. [15] | 0.1135 | 0.0970 | 0.0882 | 0.0806 | 0.0756 |
| | | | Present | 0.1134 | 0.0963 | 0.0868 | 0.0788 | 0.0740 |
| | | 0.2 | TSDT [59] | 0.4154 | 0.3606 | 0.3299 | 0.30160 | 0.2765 |
| | | | Quasi-3D [28] | 0.4168 | 0.3586 | 0.3260 | 0.29610 | 0.2722 |
| | | | Li et al. [15] | 0.4168 | 0.3586 | 0.3260 | 0.29610 | 0.2723 |
| | | | Present | 0.4150 | 0.3551 | 0.3205 | 0.28920 | 0.2667 |
| 100 | 0 | 0.1 | TSDT [59] | 0.1162 | 0.1012 | 0.0933 | 0.0867 | 0.0821 |
| | | | Quasi-3D [28] | 0.1163 | 0.1006 | 0.0923 | 0.0853 | 0.0809 |
| | | | Li et al. [15] | 0.1163 | 0.1006 | 0.0923 | 0.0853 | 0.0809 |
| | | | Present | 0.1161 | 0.0999 | 0.0910 | 0.0836 | 0.0795 |
| | | 0.2 | TSDT [59] | 0.4273 | 0.3758 | 0.3476 | 0.3219 | 0.2999 |
| | | | Quasi-3D [28] | 0.4284 | 0.3734 | 0.3431 | 0.3159 | 0.2950 |
| | | | Li et al. [15] | 0.4281 | 0.373 | 0.3427 | 0.3154 | 0.2946 |
| | | | Present | 0.4269 | 0.3702 | 0.3381 | 0.3097 | 0.2901 |
| 100 | 100 | 0.1 | TSDT [59] | 0.1619 | 0.1563 | 0.1542 | 0.1535 | 0.1543 |
| | | | Quasi-3D [28] | 0.1616 | 0.1551 | 0.1525 | 0.1512 | 0.1521 |
| | | | Li et al. [15] | 0.1613 | 0.1548 | 0.1521 | 0.1509 | 0.1517 |
| | | | Present | 0.1617 | 0.1549 | 0.1519 | 0.1505 | 0.1515 |
| | | 0.2 | TSDT [59] | 0.6162 | 0.6026 | 0.5978 | 0.5970 | 0.5993 |
| | | | Quasi-3D [28] | 0.6137 | 0.5940 | 0.5856 | 0.5815 | 0.5843 |
| | | | Li et al. [15] | 0.6085 | 0.5878 | 0.5792 | 0.5750 | 0.5720 |
| | | | Present | 0.6156 | 0.5950 | 0.5852 | 0.5810 | 0.5834 |

The results of the first four orders of non-dimensional free vibration responses based on different CNTs distribution types (FG-V, FG-O, and FG-X) and aspect ratios ($a/h$ = 10, 20 and 50) for the FG plates are given in Table 6. And it should be mentioned that the matrix is pure metal in this case. Comparing the obtained result with Selim et al. [60] and Pasha et al. [61], it can be observed that the present modeling results are in good agreement with the existing ones of Selim and Pasha. The material properties used are as follows: $E_m$ = 2.1 GPa; $\rho_m$ = 1150 kg/m$^3$; and $\mu_m = 0.34$. The efficiency parameters used are similar to those assumed by Shen [55]: $\eta_1$ = 0.149, $\eta_2$ = 0.939, and $\eta_3 = \eta_2$ for $V_{CNT}$ = 0.11.

Since there are no available literature results on the novel hybrid composite plates for comparison, the present results with different CNT volume fractions are compared with finite element ones. Moreover, Reddy shear shape function is also used for comparison. Results based on the aforementioned methodologies are shown in Table 7 (comparisons of the mode shapes are shown in Figure 4). The correctness and viability of the current model are demonstrated by the good agreement between the frequencies and other data.

**Table 6.** Validation study of non-dimensional first four free vibration frequencies of SSSS CNTRC plate based on different CNTs configurations and aspect ratios ($V_{CNT}$ = 0.11).

| Type | a/h | Mode | Method | | |
|---|---|---|---|---|---|
| | | | TSDT [60] | RPT [61] | Present |
| FG-V | 10 | (1,1) | 12.461 | 12.755 | 12.4138 |
| | | (1,2) | 17.077 | 17.128 | 16.9426 |
| | | (2,1) | 27.401 | 27.158 | 31.2429 |
| | | (2,2) | 31.943 | 33.227 | 33.558 |
| | 20 | (1,1) | 15.09 | 15.127 | 15.040 |
| | | (1,2) | 19.883 | 19.606 | 19.715 |
| | | (2,1) | 38.936 | 38.855 | 47.316 |
| | | (2,2) | 47.858 | 48.298 | 49.655 |
| | 50 | (1,1) | 16.230 | 16.093 | 16.1644 |
| | | (1,2) | 21.114 | 20.683 | 20.9135 |
| | | (2,1) | 60.094 | 59.872 | 59.4529 |
| | | (2,2) | 62.682 | 62.118 | 61.9633 |
| FG-O | 10 | (1,1) | 11.319 | 11.773 | 9.568 |
| | | (1,2) | 16.137 | 16.469 | 14.906 |
| | | (2,1) | 26.555 | 26.827 | 26.553 |
| | | (2,2) | 29.408 | 31.858 | 29.507 |
| | 20 | (1,1) | 13.405 | 13.50 | 10.567 |
| | | (1,2) | 18.398 | 18.371 | 16.309 |
| | | (2,1) | 38.938 | 38.855 | 34.945 |
| | | (2,2) | 42.997 | 44.759 | 38.272 |
| | 50 | (1,1) | 14.265 | 14.153 | 10.914 |
| | | (1,2) | 19.338 | 19.154 | 16.803 |
| | | (2,1) | 52.653 | 52.616 | 39.268 |
| | | (2,2) | 55.451 | 55.123 | 42.846 |
| FG-X | 10 | (1,1) | 14.683 | 15.254 | 14.586 |
| | | (1,2) | 18.687 | 18.825 | 18.551 |
| | | (2,1) | 28.524 | 28.004 | 33.447 |
| | | (2,2) | 34.442 | 35.980 | 35.488 |
| | 20 | (1,1) | 19.94 | 20.241 | 19.866 |
| | | (1,2) | 23.771 | 23.573 | 23.611 |
| | | (2,1) | 38.938 | 38.855 | 56.591 |
| | | (2,2) | 52.336 | 49.581 | 58.344 |
| | 50 | (1,1) | 22.954 | 22.880 | 22.873 |
| | | (1,2) | 26.741 | 26.183 | 26.547 |
| | | (2,1) | 83.153 | 83.604 | 82.218 |
| | | (2,2) | 84.956 | 83.703 | 83.992 |

**Table 7.** Comparison of the first eight natural frequencies (Hz) of three-dimensional sandwich plates under different CNT volume fractions with simple support conditions.

| $V_{CNT}$ | Method | Mode | | | | | | | | |
|---|---|---|---|---|---|---|---|---|---|---|
| | | 1 | 2 | 3 | 4 | 5 | 6 | 7 | 8 | *Diff.* (%) |
| 0.12 | ABAQUS | 668.7 | 1552.9 | 1698.1 | 2387.5 | 2890.9 | 3188.7 | 3552.7 | 3725.4 | – |
| | Present—Reddy [62] | 669.0 | 1554.0 | 1700.0 | 2391.9 | 2895.3 | 3195.9 | 3560.4 | 3734.8 | 0.16 |
| | Present | 668.7 | 1555.0 | 1701.6 | 2391.8 | 2899.6 | 3201.1 | 3561.8 | 3737.1 | 0.22 |
| 0.17 | ABAQUS | 835.5 | 2012.6 | 2020.8 | 3004.6 | 3765.2 | 3781.6 | 4577.8 | 4589.9 | – |
| | Present—Reddy [62] | 834.8 | 2012.6 | 2018.8 | 3006.1 | 3767.9 | 3777.6 | 4582.5 | 4590.2 | 0.06 |
| | Present | 832.8 | 2011.6 | 2017.6 | 3001.4 | 3771.0 | 3780.6 | 4580.0 | 4587.5 | 0.11 |
| 0.28 | ABAQUS | 922.9 | 2193.7 | 2269.2 | 3308.1 | 4095.9 | 4254.8 | 5001.3 | 5089.4 | – |
| | Present—Reddy [62] | 916.3 | 2178.6 | 2256.7 | 3291.4 | 4067.0 | 4235.7 | 4973.2 | 5067.9 | 0.57 |
| | Present | 914.7 | 2181.3 | 2260.6 | 3290.8 | 4081.4 | 4252.5 | 4979.5 | 5076.5 | 0.43 |

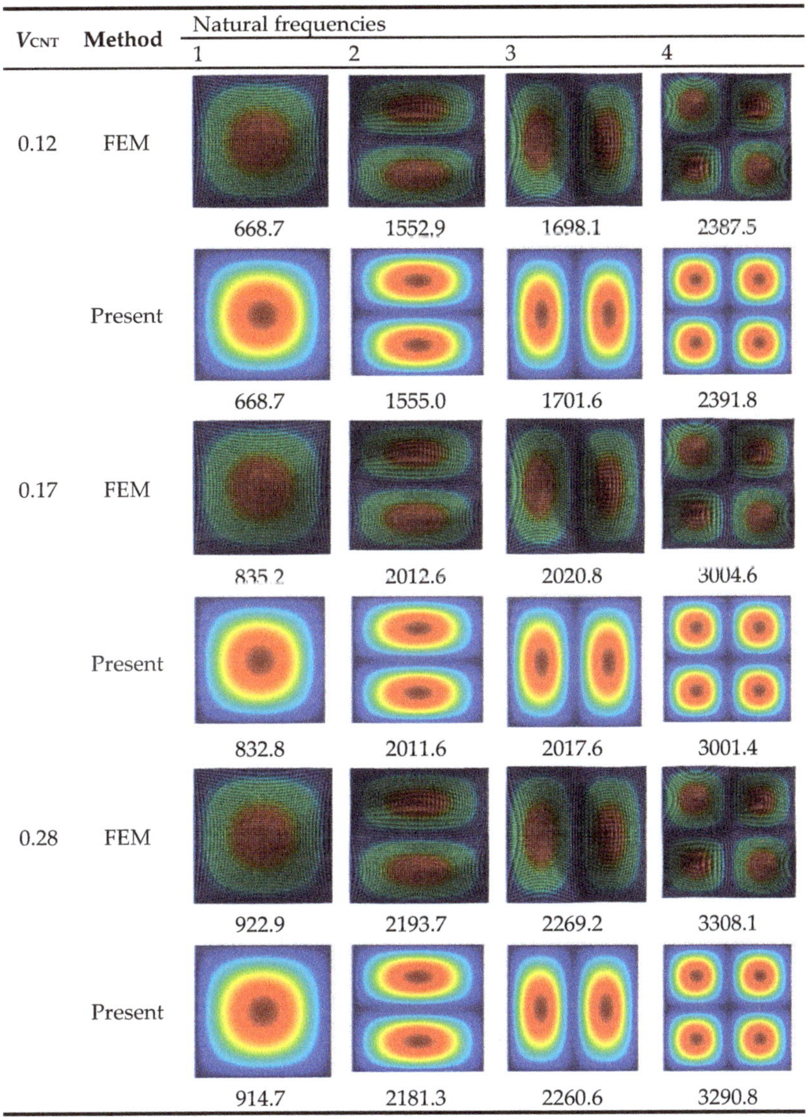


**Figure 4.** Comparison of the first four mode shapes of sandwich plates under different CNT volume fractions with simple support conditions.

### 3.2. Parametric Studies

This section examines how the free vibration frequencies of sandwich plates set on Winkler–Pasternak foundations are affected by geometric parameters, material characteristics, and elastic stiffnesses. Figure 5 demonstrates the effect of the thickness ratio $h_c/h$, the elasticity modulus ratios $E_1/E_2$, and gradient index $p$ on the dimensionless fundamental frequency of the sandwich plate. The thickness of sandwich plate is $h$ = 0.1 m. Figure 5 illustrates that when the elastic modulus ratio rises, so does the dimensionless fundamental frequency. However, as Figure 5 illustrates, when the $h_c/h$ increases from 0.2 to 0.5, the dimensionless fundamental frequencies somewhat drop for $p$ = 1, 5, and 10 and $E_1/E_2$ = 5. Furthermore, when the gradient index rises, the dimensionless fundamental frequency falls, with the decline being more rapid for large $h_c/h$. Figure 5 similarly shows a sharp rise in the dimensionless fundamental frequency as thickness ratios grow.

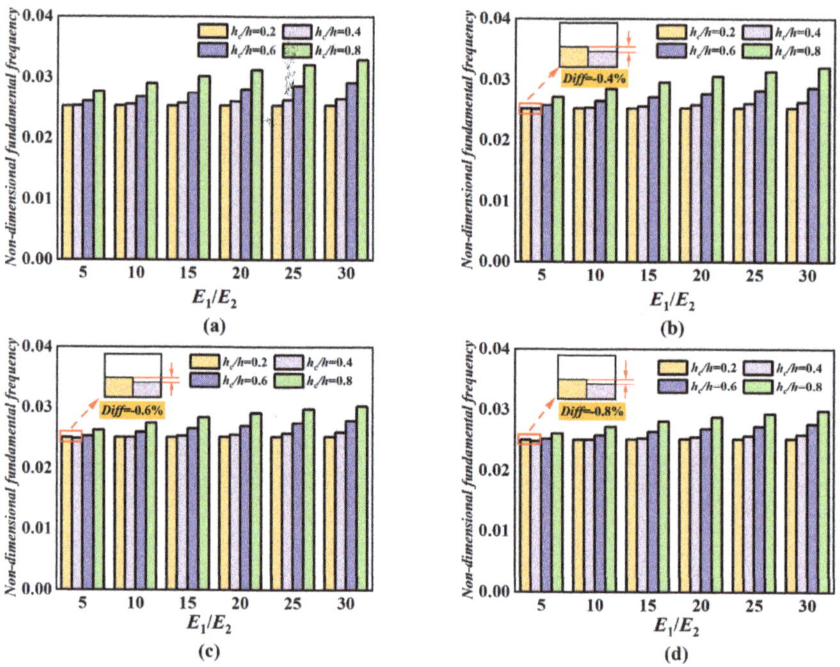

**Figure 5.** The variation in natural frequency with Young's modulus ratio ($E_1/E_2$) under the ratio of core thickness to boundary thickness ($h_c/h$). (**a**) $p$ = 0.5, (**b**) $p$ = 1, (**c**) $p$ = 5, and (**d**) $p$ = 10.

Figure 6 examines the impact of the gradient index ($p$ = 0.5, 1, 5, and 10), core thickness ratio, and CNT volume fraction ($V_{CNT}$ = 0, 0.12, 0.17, and 0.28) on the dimensionless fundamental frequency of sandwich plates. It is evident that the sandwich plate's dimensionless fundamental frequency grows dramatically as the core thickness ratio rises from 0.2 to 1, while it remains nearly constant for core thickness ratios between 0 and 0.2. Furthermore, as the volume percentage of CNTs grows, so does the dimensionless fundamental frequency. It has been demonstrated that the structural rigidity of sandwich plates can be considerably increased by increasing their thickness and CNT volume fraction.

Figure 7 shows how the dimensionless fundamental frequency of sandwich plates varies with changes in the gradient index, core thickness ratio, and elastic stiffness parameters. The sandwich plate is 0.1 m thick. This image illustrates how the fundamental frequency rapidly decreases for small graded indices and decreases with an increase in the gradient index $p$. The reason behind this phenomenon is that the elastic modulus decease with the increase in graded indexes, and this leads to the decrease in bending stiffness of the whole sandwich plates. In addition, the dimensionless fundamental frequency significantly increases with core thicknesses, indicating that the hybrid composite core can significantly improve the structural stiffness. Additionally, Figure 7 shows that the dimensionless fundamental frequency of the sandwich plate rises as $kw$ and $Gp$ grow, but that $kw$ has little effect on the sandwich plate's vibration frequency. And the dimensionless fundamental frequency of the sandwich plate is greatly impacted by the increase in $Gp$. The sandwich plate's dimensionless fundamental frequency rises noticeably with increasing $Gp$. Furthermore, thinner core plates exhibit larger dimensionless fundamental frequencies at $Gp$ = 100 compared to $Gp$ = 0 for lower gradient indexes.

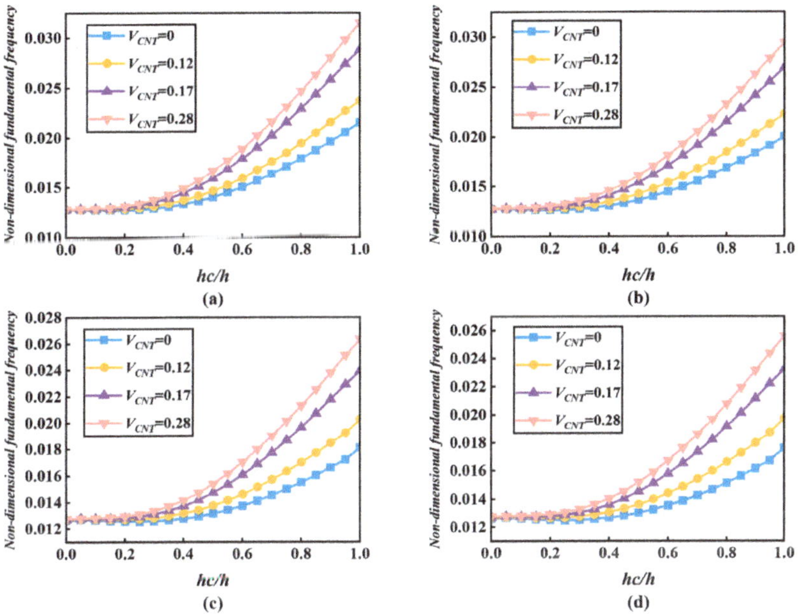


**Figure 6.** The impact of CNT volume percentage and core thickness ratio on the hybrid sandwich plate's natural frequency. (**a**) $p = 0.5$, (**b**) $p = 1$, (**c**) $p = 5$, and (**d**) $p = 10$.

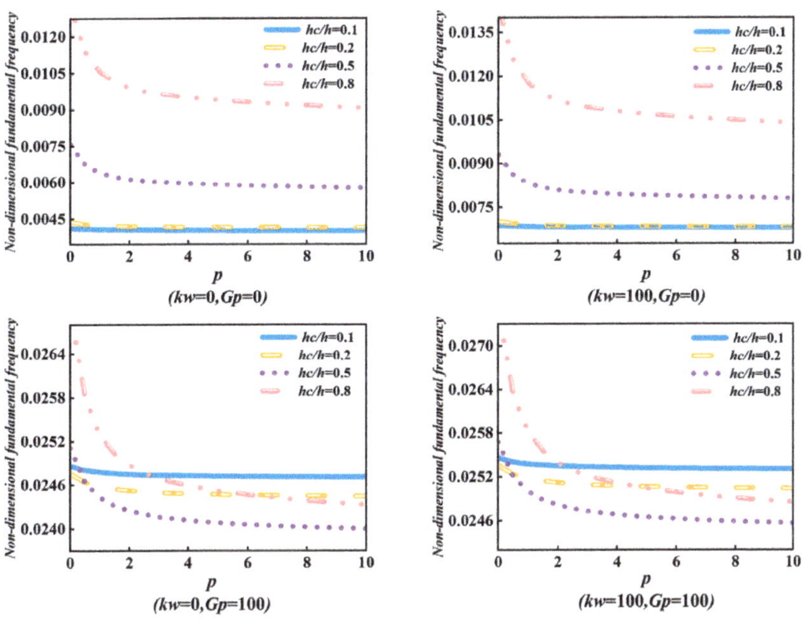


**Figure 7.** The natural frequency of three-dimensional sandwich plates and shells is affected by the gradient index, core thickness ratio, and Winkler–Pasternak foundation parameters.

Figure 8 examines the effects of gradient index, elastic parameters, and thickness-to-width ratio on the sandwich plate's dimensionless fundamental frequency. The ratio of thickness is $h_c/h = 0.8$. Because of the thin plates, the bending stiffness and fundamental frequencies rise with overall thickness, as shown in Figure 8, which shows that the dimensionless fundamental frequency steadily increases as the thickness to width ratio rises from 0.05 to 0.2. Additionally, Figure 8 shows that as *kw* and *Gp* grow, the sandwich plate's vibration frequency rises, but *kw* has less of an impact on the sandwich plate's dimensionless fundamental frequency. Additionally, as *Gp* increases, the dimensionless fundamental frequency will rise noticeably.

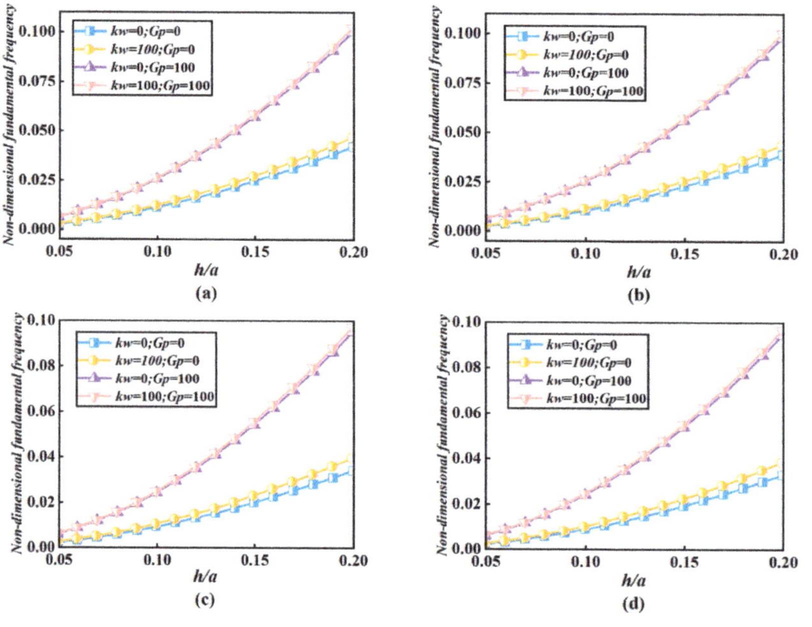


**Figure 8.** The influence of aspect ratio and Winkler–Pasternak foundation parameters on the natural frequency of three-dimensional sandwich plates and shells. (**a**) $p = 0.5$, (**b**) $p = 1$, (**c**) $p = 5$, and (**d**) $p = 10$.

The dimensionless fundamental frequency of sandwich plates with varying aspect ratios is examined in Table 8. Additionally, the impact of the volume percentage and distribution type of carbon nanotubes is taken into account. The dimensionless fundamental frequency is found to drop as the aspect ratio rises, with the tendency for the reduction to slow down as the aspect ratio approaches reasonably large values. For example, for the FG-V distribution with $kw = Gp = 100$, the dimensionless frequency decreases by 55.68% when the aspect ratio is increased from $a/b = 1$ to $a/b = 2$, and the dimensionless fundamental frequency decreases by 25.61% when $a/b$ is increased from 2 to 3. This trend indicates that the increase in the volume fraction of carbon nanotubes can enhance the structural stiffness. Furthermore, the distribution of CNTs along the thickness direction significantly influences the free vibration of the sandwich plate because the FG-X has the highest dimensionless fundamental frequency compared to other distribution types.

Figure 9 examines the effects of CNT distribution types, volume fraction, and gradient index on the sandwich plate's dimensionless fundamental frequency. Figure 9 shows that when CNTs have distinct distribution types in the core, FG-O has a relatively modest dimensionless fundamental frequency and FG-X is slightly greater than the other types. With the exception of the FG-O distribution type, it can be seen that the three CNT distribution types—UD, FG-V, and FG-X—have no discernible impact on the dimensionless fundamental frequency. Furthermore, Figure 9 demonstrates that as the volume proportion of CNTs grows, so does the dimensionless fundamental frequency. However, when the gradient index progressively rises, the dimensionless fundamental frequency falls.

**Table 8.** Simple supported hybrid sandwich plates with different aspect ratios ($h_c$ = 0.1; $h_f$ = 0.01) for the variation in dimensionless natural frequencies ($p$ = 1).

| (kw,Gp) | $V_{CNT}$ | | a/b | | | | | | | |
|---|---|---|---|---|---|---|---|---|---|---|
| | | | 1 | 1.2 | 1.4 | 1.6 | 1.8 | 2 | 3 | 4 |
| (0,0) | 0.12 | UD | 0.0158 | 0.0133 | 0.0118 | 0.0109 | 0.0103 | 0.0099 | 0.0089 | 0.0086 |
| | | FG-V | 0.0155 | 0.0131 | 0.0117 | 0.0108 | 0.0103 | 0.0099 | 0.0089 | 0.0086 |
| | | FG-X | 0.0164 | 0.0138 | 0.0122 | 0.0112 | 0.0106 | 0.0101 | 0.0092 | 0.0088 |
| | | FG-O | 0.0145 | 0.0121 | 0.0111 | 0.0103 | 0.0097 | 0.0094 | 0.0085 | 0.0082 |
| | 0.17 | UD | 0.0196 | 0.0168 | 0.0151 | 0.0140 | 0.0133 | 0.0128 | 0.0116 | 0.0112 |
| | | FG-V | 0.0193 | 0.0166 | 0.0150 | 0.0140 | 0.0133 | 0.0129 | 0.0117 | 0.0113 |
| | | FG-X | 0.0207 | 0.0176 | 0.0149 | 0.0147 | 0.0139 | 0.0134 | 0.0121 | 0.0117 |
| | | FG-O | 0.0177 | 0.0153 | 0.0139 | 0.0130 | 0.0124 | 0.0119 | 0.0109 | 0.0105 |
| | 0.28 | UD | 0.0216 | 0.0183 | 0.0164 | 0.0152 | 0.0144 | 0.0138 | 0.0125 | 0.0121 |
| | | FG-V | 0.0212 | 0.0182 | 0.0164 | 0.0153 | 0.0146 | 0.014 | 0.0128 | 0.0123 |
| | | FG-X | 0.0237 | 0.0201 | 0.018 | 0.0167 | 0.0158 | 0.0152 | 0.0138 | 0.0133 |
| | | FG-O | 0.0182 | 0.0157 | 0.0142 | 0.0132 | 0.0125 | 0.0121 | 0.0110 | 0.0107 |
| (100,100) | 0.12 | UD | 0.0372 | 0.029 | 0.0239 | 0.0206 | 0.0182 | 0.0165 | 0.0122 | 0.0105 |
| | | FG-V | 0.0370 | 0.0289 | 0.0239 | 0.0205 | 0.0182 | 0.0164 | 0.0122 | 0.0105 |
| | | FG-X | 0.0375 | 0.0292 | 0.0241 | 0.0208 | 0.0184 | 0.0166 | 0.0123 | 0.0107 |
| | | FG-O | 0.0363 | 0.0283 | 0.0233 | 0.0200 | 0.0177 | 0.0160 | 0.0118 | 0.0101 |
| | 0.17 | UD | 0.0394 | 0.0310 | 0.0259 | 0.0226 | 0.0202 | 0.0185 | 0.0143 | 0.0128 |
| | | FG-V | 0.0392 | 0.0309 | 0.0259 | 0.0225 | 0.0202 | 0.0185 | 0.0144 | 0.0129 |
| | | FG-X | 0.0399 | 0.0315 | 0.0264 | 0.0230 | 0.0206 | 0.0189 | 0.0147 | 0.0132 |
| | | FG-O | 0.0378 | 0.0298 | 0.0249 | 0.0216 | 0.0193 | 0.0177 | 0.0136 | 0.0121 |
| | 0.28 | UD | 0.0412 | 0.0325 | 0.0272 | 0.0237 | 0.0213 | 0.0195 | 0.0152 | 0.0136 |
| | | FG-V | 0.0409 | 0.0324 | 0.0272 | 0.0237 | 0.0213 | 0.0196 | 0.0154 | 0.0139 |
| | | FG-X | 0.0424 | 0.0336 | 0.0282 | 0.0247 | 0.0223 | 0.0205 | 0.0163 | 0.0147 |
| | | FG-O | 0.0384 | 0.0303 | 0.0253 | 0.0219 | 0.0196 | 0.0179 | 0.0138 | 0.0123 |

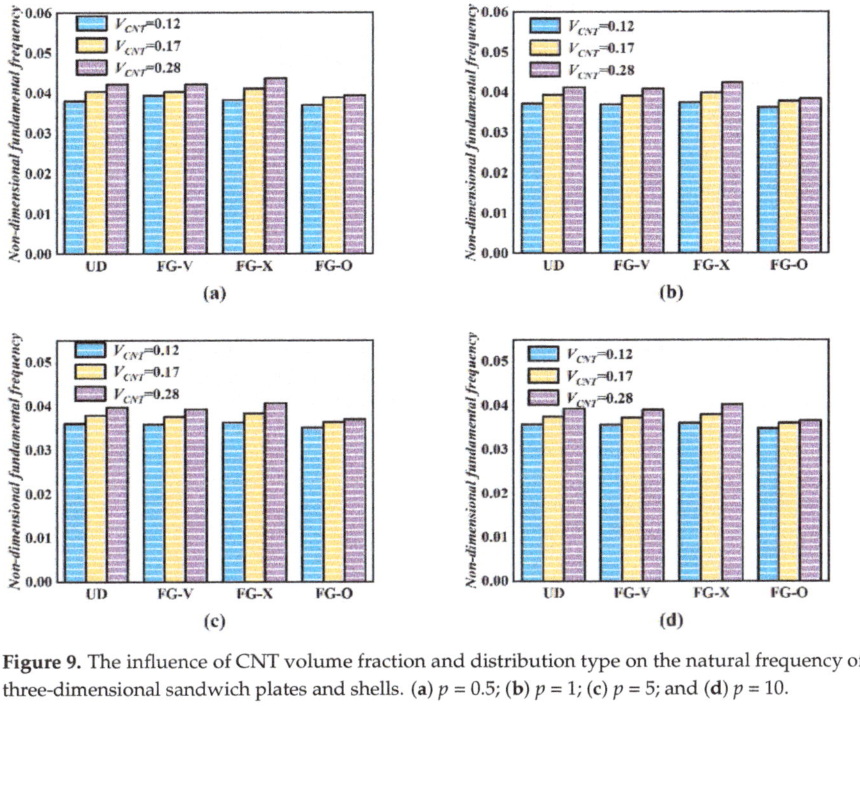


**Figure 9.** The influence of CNT volume fraction and distribution type on the natural frequency of three-dimensional sandwich plates and shells. **(a)** $p$ = 0.5; **(b)** $p$ = 1; **(c)** $p$ = 5; and **(d)** $p$ = 10.

The effect of elastic foundation parameters on the sandwich plate's dimensionless fundamental frequency is investigated in Figure 10. The distribution types and volume proportion of CNTs are taken into account. Generally speaking, when the volume percentage of CNTs rises, the dimensionless fundamental frequency rises noticeably, demonstrating that CNTs increase the material's rigidity. It is also observed that the dimensionless fundamental frequency increases with an increase in *kw* and *Gp*. However, the impact of *kw* on the dimensionless fundamental frequency is less, and the increase in *Gp* significantly increases the dimensionless fundamental frequency. Moreover, in good agreement with expectation, among the four distributions, FG-O has the relatively smallest dimensionless fundamental frequency and FG-X has the relatively largest dimensionless fundamental frequency.

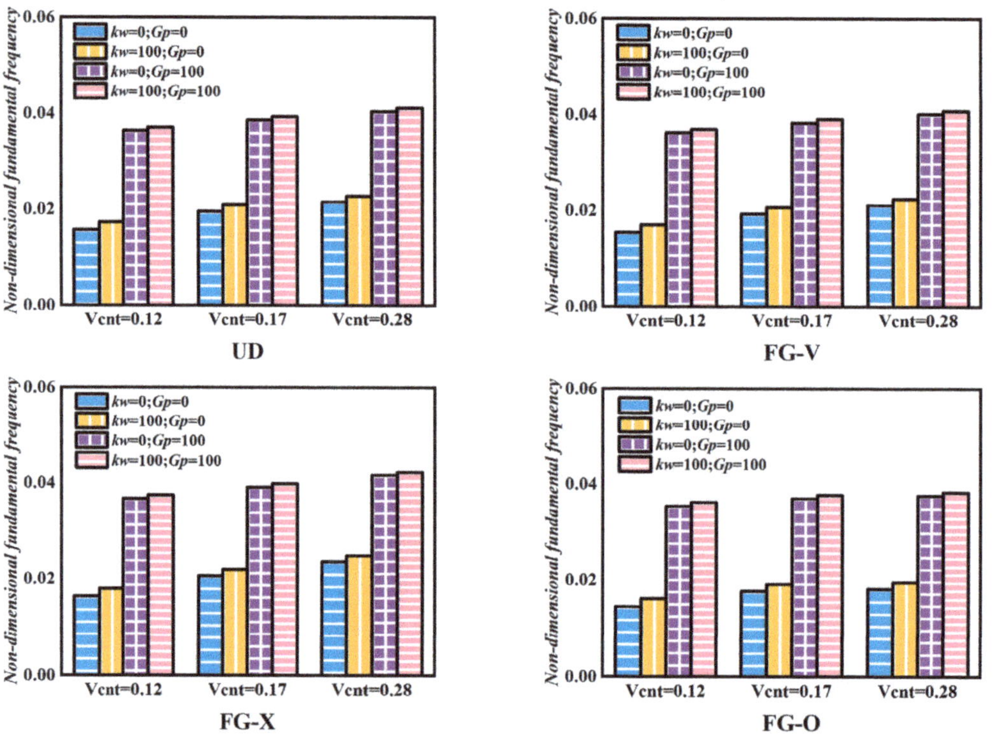


**Figure 10.** The influence of CNT volume fraction and Winkler–Pasternak foundation parameters on the free vibration of three-dimensional sandwich plates under different CNT distribution types.

Figure 11 investigates the effect of fiber's various lamination schemes and elastic foundation parameters of the plate on the first four dimensionless free vibration frequencies of the sandwich plate. It should be noted that the influence of different lamination schemes has little effect on the first two dimensionless frequencies, but as the order increases, the differences in the dimensionless frequencies of the different lamination schemes become progressively more pronounced. In the last two free vibration frequencies, the dimensionless frequency is highest when the lamination scheme of the fiber is $[0°/90°]_{4s}$ and is lowest when the lamination scheme of the fiber is $[0°]_{4s}$. In addition, in agreement with the previous cases, the dimensionless frequencies significantly increase with the increase in the elastic foundation parameters *kw* and *Gp*.

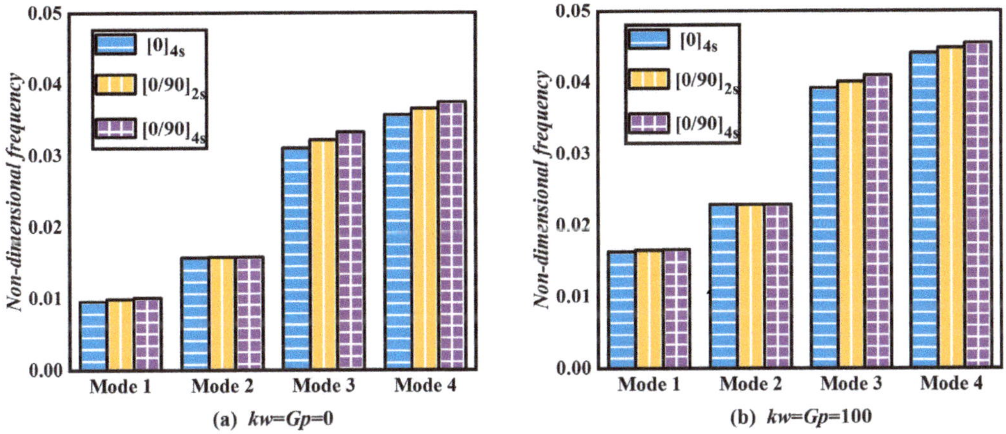


**Figure 11.** Effect of fiber's lamination schemes and Winkler–Pasternak basis parameters on the first four orders of dimensionless frequency of three-dimensional sandwich plates.

## 4. Conclusions

The free vibration behavior of a sandwich composite plate on the Winkler/Pasternak elastic basis is investigated analytically in this work. The sheets are a fiber-reinforced composite, whereas the core is a CNT-reinforced composite with a metal–ceramic FGM matrix. In the FGM plate, the CNTs are arranged in four different configurations. The HSDT-based motion equations are derived through the Hamilton energy principle. The free vibration solutions are then obtained by applying the Navier method. Furthermore, the accuracy of this research approach is confirmed by comparing the results of this paper's method with those of finite elements and previous studies using several examples. The structure studied in this paper can not only promote the application of composites in the field of vibration control but also provide an important theoretical basis and technical support for improving the performance, reliability, and multifunctionality of engineering structures. In addition, the effect of characteristics parameters on the dimensionless frequency of sandwich plates is investigated. And some results of the study are observed as follows:

(1) As the volume fraction of ceramics and CNTs increases, it leads to an increase in the stiffness of the plate and produces larger dimensionless free vibration frequencies.
(2) For different distributions of CNT (UD, FG-V, FG-X, FG-O), the FG-O distribution significantly reduces the dimensionless frequency, while the remaining three distributions have no significant effect on the dimensionless frequency.
(3) Both the elastic foundation parameters *kw* and *Gp* have positively correlated effects on the free vibration behavior of the FRP_FG-CNTRC_FRP plate. However, *Gp* affects the dimensionless fundamental frequency to a much greater extent than *kw*.
(4) It can be clearly observed that the dimensionless fundamental frequency of the sandwich composite plate increases significantly with the increase in $h_c/h$, indicating that the FG-CNTRC core layer mentioned in this paper can enhance the stiffness of the plate well.

**Author Contributions:** M.L., manuscript writing, conceptualization, study design, and supervision; X.L., manuscript writing, data curation, and methodology; M.Y., literature search and validation; W.C., supervision and methodology. All authors have read and agreed to the published version of the manuscript.

**Funding:** The research work reported in this paper is supported by the National Natural Science Foundation of China (project 12202324).

**Institutional Review Board Statement:** Not applicable.

**Informed Consent Statement:** Not applicable.

**Data Availability Statement:** Data are contained within this article.

**Conflicts of Interest:** The authors declare no conflicts of interest.

## Appendix A

$$\begin{aligned}\delta u_0: & A_{11}d_{11}u_0 - B_{11}d_{111}w_0 + C_{11}d_{11}\theta_x + A_{12}d_{12}v_0 - B_{12}d_{122}w_0 + C_{12}d_{12}\theta_y + A_{16}(d_{12}u_0 + d_{11}v_0) \\ & -2B_{16}d_{112}w_0 + C_{16}(d_{12}\theta_x + d_{11}\theta_y) + A_{16}d_{12}u_0 - B_{16}d_{112}w_0 + C_{16}d_{12}\theta_x + A_{26}d_{22}v_0 - B_{26}d_{222}w_0 \\ & +C_{26}d_{22}\theta_y + A_{66}(d_{22}u_0 + d_{12}v_0) - 2B_{66}d_{122}w_0 + C_{66}(d_{22}\theta_x + d_{12}\theta_y) = I_0\ddot{u}_0 - I_1d_1\ddot{w}_0 - I_3d_1\ddot{\theta}_x\end{aligned} \tag{A1}$$

$$\begin{aligned}\delta v_0: & A_{12}d_{12}u_0 - B_{12}d_{112}w_0 + C_{12}d_{12}\theta_x + A_{22}d_{22}v_0 - B_{22}d_{222}w_0 + C_{22}d_{22}\theta_y + A_{26}(d_{22}u_0 + d_{12}v_0) \\ & -2B_{26}d_{122}w_0 + C_{26}(d_{22}\theta_x + d_{12}\theta_y) + A_{16}d_{11}u_0 - B_{16}d_{1111}w_0 + C_{16}d_{11}\theta_x + A_{26}d_{12}v_0 - B_{26}d_{122}w_0 \\ & +C_{26}d_{12}\theta_y + A_{66}(d_{12}u_0 + d_{11}v_0) - 2B_{66}d_{112}w_0 + C_{66}(d_{12}\theta_x + d_{11}\theta_y) = I_0\ddot{v}_0 - I_1d_2\ddot{w}_0 - I_3d_2\ddot{\theta}_y\end{aligned} \tag{A2}$$

$$\begin{aligned}\delta w_0: & B_{11}d_{111}u_0 - D_{11}d_{1111}w_0 + E_{11}d_{111}\theta_x + B_{12}d_{112}v_0 - D_{12}d_{1122}w_0 + E_{12}d_{112}\theta_y + B_{16}(d_{112}u_0 + d_{111}v_0) \\ & -2D_{16}d_{1112}w_0 + E_{16}(d_{112}\theta_x + d_{111}\theta_y) + B_{12}d_{122}u_0 - D_{12}d_{1122}w_0 + E_{12}d_{122}\theta_x + B_{22}d_{222}v_0 - D_{22}d_{2222}w_0 \\ & +E_{22}d_{222}\theta_y + B_{26}(d_{222}u_0 + d_{122}v_0) - 2D_{26}d_{1222}w_0 + E_{26}(d_{222}\theta_x + d_{122}\theta_y) + 2B_{16}d_{112}u_0 - 2D_{16}d_{1112}w_0 \\ & +2E_{16}d_{112}\theta_x + 2B_{26}d_{122}u_0 - 2D_{26}d_{1222}w_0 + 2E_{26}d_{122}\theta_y + 2B_{66}(d_{122}u_0 + d_{112}v_0 - 4D_{66}d_{1122}w_0 \\ & +2E_{66}(d_{122}\theta_x + d_{112}\theta_y) \underbrace{- \underbrace{k_w w_0}_{q_{Winker}} + G_p\nabla^2 w_0}_{q_{Pasternak}} = I_0\ddot{w}_0 + I_1(d_1\ddot{u}_0 + d_2\ddot{v}_0) - I_4(d_{11}\ddot{\theta}_x + d_{22}\ddot{\theta}_y)\end{aligned} \tag{A3}$$

$$\begin{aligned}\delta\theta_x: & C_{11}d_{11}u_0 - E_{11}d_{111}w_0 + F_{11}d_{22}\theta_x + C_{12}d_{12}v_0 - E_{12}d_{122}w_0 + F_{12}d_{12}\theta_y + C_{16}(d_{12}u_0 + d_{11}v_0) \\ & -2E_{16}d_{112}w_0 + F_{16}(d_{12}\theta_x + d_{11}\theta_y) + C_{16}d_{12}u_0 - E_{16}d_{112}w_0 + F_{16}d_{12}\theta_x + C_{26}d_{22}v_0 - E_{26}d_{222}w_0 \\ & +F_{26}d_{22}\theta_y + C_{66}(d_{22}u_0 + d_{12}v_0) - 2E_{66}d_{122}w_0 + F_{66}(d_{22}\theta_x + d_{12}\theta_y) - G_{45}\theta_y - G_{55}\theta_x \\ & = I_2\ddot{u}_0 - I_4d_1\ddot{w}_0 + I_5\ddot{\theta}_x\end{aligned} \tag{A4}$$

$$\begin{aligned}\delta\theta_y: & C_{12}d_{12}u_0 - D_{12}d_{112}w_0 + F_{12}d_{12}\theta_x + C_{22}d_{22}v_0 - D_{22}d_{222}w_0 + F_{22}d_{22}\theta_y + C_{26}(d_{22}u_0 + d_{12}v_0) \\ & -2D_{26}d_{122}w_0 + F_{26}(d_{22}\theta_x + d_{12}\theta_y) + C_{16}d_{11}u_0 - E_{16}d_{111}w_0 + F_{16}d_{11}\theta_x + C_{26}d_{12}v_0 - E_{26}d_{122}w_0 \\ & +F_{26}d_{12}\theta_y + C_{66}(d_{12}u_0 + d_{22}v_0) - 2E_{66}d_{112}w_0 + F_{66}(d_{12}\theta_x + d_{11}\theta_y) - G_{44}\theta_y - G_{45}\theta_x \\ & = I_2\ddot{v}_0 - I_4d_2\ddot{w}_0 + I_5\ddot{\theta}_y\end{aligned} \tag{A5}$$

where $d_i$,$d_{ij}$,$d_{ijm}$, and $d_{ijmn}$ are the symbols for the differential operations, which have the following specific expression forms:

$$\begin{aligned} & d_i = \frac{\partial}{\partial\lambda_i}, d_{ij} = \frac{\partial}{\partial\lambda_i\partial\lambda_j}, d_{ijm} = \frac{\partial}{\partial\lambda_i\partial\lambda_j\partial\lambda_m}, d_{ijmn} = \frac{\partial}{\partial\lambda_i\partial\lambda_j\partial\lambda_m\partial\lambda_n}, (i,j,m,n = 1,2) \\ & \lambda = x, (i,j,m,n = 1), \lambda = y, (i,j,m,n = 2)\end{aligned} \tag{A6}$$

## References


1. Li, M.; Yan, R.; Shen, W.; Qin, K.; Li, J.; Liu, K. Fatigue Characteristics of Sandwich Composite Joints in Ships. *Ocean Eng.* **2022**, *254*, 111254. [CrossRef]
2. Li, M.; Liu, Z.; Yan, R.; Lu, J.; Guedes Soares, C. Experimental and Numerical Investigation on Composite Single-Lap Single-Bolt Sandwich Joints with Different Geometric Parameters. *Mar. Struct.* **2022**, *85*, 103259. [CrossRef]
3. Wang, H.; Li, M.; Liu, X. A Non-Uniform Equivalent Model for Free Vibration Analysis of Sandwich Composite Panels with Trapezoidal Lattice Core. *J. Vib. Eng. Technol.* **2024**, *12*, 7009–7019. [CrossRef]
4. Qiu, Y.; Shen, W.; Yan, R.; Li, X.; Ye, Z.; Li, M.; Liu, K.; Qin, K. An Improved Numerical Method for Calculating Mechanical Properties of Bi-Modulus Sandwich Composite Structures. *Ocean Eng.* **2022**, *250*, 110998. [CrossRef]
5. Meng, S.; Zhong, R.; Wang, Q.; Shi, X.; Qin, B. Vibration Characteristic Analysis of Three-Dimensional Sandwich Cylindrical Shell Based on the Spectro-Geometric Method. *Compos. Struct.* **2024**, *327*, 117661. [CrossRef]
6. Hu, S.; Wang, Q.; Zhong, R.; Qin, B. Random Thermal-Vibration Mechanisms of Sandwich Ventral Fin-Type Plate-Shell Systems with Porous Functionally Graded Core. *Thin-Walled Struct.* **2024**, *204*, 112333. [CrossRef]
7. Chen, W.; Chen, C.; Zhang, Y.; Li, P.; Li, M.; Li, X. Study on the Deformation Mode and Energy Absorption Characteristics of Protective Honeycomb Sandwich Structures Based on the Combined Design of Lotus Root Nodes and Leaf Stem Veins. *J. Mar. Sci. Eng.* **2024**, *12*, 652. [CrossRef]

8. Fan, Z.; Li, X.; Huang, T.; Chen, W. An Investigation of the Damage Mechanism of Multilayer Liquid-Containing Protective Structure under Combined Blast Wave and Fragment Loading. *J. Mar. Sci. Eng.* **2023**, *11*, 2327. [CrossRef]
9. Corigliano, P.; Palomba, G.; Crupi, V.; Garbatov, Y. Stress-Strain Assessment of Honeycomb Sandwich Panel Subjected to Uniaxial Compressive Load. *J. Mar. Sci. Eng.* **2023**, *11*, 365. [CrossRef]
10. Lin, H.; Han, C.; Yang, L.; Zhang, L.; Luan, H.; Han, P.; Xu, H.; Zhang, S. Numerical Investigation on Performance Optimization of Offshore Sandwich Blast Walls with Different Honeycomb Cores Subjected to Blast Loading. *J. Mar. Sci. Eng.* **2022**, *10*, 1743. [CrossRef]
11. Chen, H.; Cai, Y.; Zhang, J.; Lv, X.; Li, X. Analytical Solutions for Out-of-Plane Response of Curved Beams Resting on an Elastic Foundation under a Random Moving Load. *Eng. Struct.* **2024**, *318*, 118753. [CrossRef]
12. Rui, S.; Zhang, H.; Xu, H.; Zha, X.; Xu, M.; Shen, K. Seabed Structures and Foundations Related to Deep-Sea Resource Development: A Review Based on Design and Research. *Deep Undergr. Sci. Eng.* **2024**, *3*, 131–148. [CrossRef]
13. Li, M.; Yan, R.; Guedes Soares, C. Free Vibration of Advanced Composite Plates Using a New Higher Order Shear Deformation Theory. *Eur. J. Mech.-A/Solids* **2021**, *88*, 104236. [CrossRef]
14. Li, M.; Yan, R.; Xu, L.; Guedes Soares, C. A General Framework of Higher-Order Shear Deformation Theories with a Novel Unified Plate Model for Composite Laminated and FGM Plates. *Compos. Struct.* **2021**, *261*, 113560. [CrossRef]
15. Li, M.; Guedes Soares, C.; Yan, R. Free Vibration Analysis of FGM Plates on Winkler/Pasternak/Kerr Foundation by Using a Simple Quasi-3D HSDT. *Compos. Struct.* **2021**, *264*, 113643. [CrossRef]
16. Li, M.; Guedes Soares, C.; Yan, R. A Novel Shear Deformation Theory for Static Analysis of Functionally Graded Plates. *Compos. Struct.* **2020**, *250*, 112559. [CrossRef]
17. Fotouhi, S.; Fotouhi, M.; Pavlovic, A.; Djordjevic, N. Investigating the Pre-Damaged PZT Sensors under Impact Traction. *J. Mar. Sci. Eng.* **2018**, *6*, 142. [CrossRef]
18. Strozzi, M.; Giacomobono, R.; Rubini, R.; Cocconcelli, M. Preliminary Orthotropic Elastic Model for the Study of Natural Frequencies and Mode Shapes of a 3D Printed Onyx Thin Circular Cylindrical Shell. *Int. J. Mech. Control* **2020**, *21*, 51–56.
19. Arani, H.K.; Shariyat, M. Nonlinear 2D-DQ Volume-Preservative Global–Local Dynamic Analysis of Composite Sandwich Plates with Soft Hyperelastic Cores and Viscoelastic Winkler-Pasternak Foundations. *Structures* **2023**, *55*, 727–746. [CrossRef]
20. Kumar, P.; Harsha, S.P. Static Analysis of Porous Core Functionally Graded Piezoelectric (PCFGP) Sandwich Plate Resting on the Winkler/Pasternak/Kerr Foundation under Thermo-Electric Effect. *Mater. Today Commun.* **2022**, *32*, 103929. [CrossRef]
21. Hadji, L.; Avcar, M.; Zouatnia, N. Natural Frequency Analysis of Imperfect FG Sandwich Plates Resting on Winkler-Pasternak Foundation. *Polym. Mediterr. Fiber Int.* **2022**, *53*, 153–160. [CrossRef]
22. Li, Q.; Wu, D.; Chen, X.; Liu, L.; Yu, Y.; Gao, W. Nonlinear Vibration and Dynamic Buckling Analyses of Sandwich Functionally Graded Porous Plate with Graphene Platelet Reinforcement Resting on Winkler–Pasternak Elastic Foundation. *Int. J. Mech. Sci.* **2018**, *148*, 596–610. [CrossRef]
23. El-Shahrany, H.D.; Zenkour, A.M. Control of Dynamic Response of the Functionally Graded Smart Sandwich Beam Coupled Variable Kelvin–Voigt–Pasternak's Model. *Ain Shams Eng. J.* **2024**, *15*, 102476. [CrossRef]
24. Selim, B.A.; Liu, Z. Impact Analysis of Functionally-Graded Graphene Nanoplatelets-Reinforced Composite Plates Laying on Winkler-Pasternak Elastic Foundations Applying a Meshless Approach. *Eng. Struct.* **2021**, *241*, 112453. [CrossRef]
25. Zaitoun, M.W.; Chikh, A.; Tounsi, A.; Al-Osta, M.A.; Sharif, A.; Al-Dulaijan, S.U.; Al-Zahrani, M.M. Influence of the Visco-Pasternak Foundation Parameters on the Buckling Behavior of a Sandwich Functional Graded Ceramic–Metal Plate in a Hygrothermal Environment. *Thin-Walled Struct.* **2022**, *170*, 108549. [CrossRef]
26. Zenkour, A.M.; El-Shahrany, H.D. Hygrothermal Effect on Vibration of Magnetostrictive Viscoelastic Sandwich Plates Supported by Pasternak's Foundations. *Thin-Walled Struct.* **2020**, *157*, 107007. [CrossRef]
27. Singh, S.J.; Harsha, S.P. Nonlinear Dynamic Analysis of Sandwich S-FGM Plate Resting on Pasternak Foundation under Thermal Environment. *Eur. J. Mech.-A/Solids* **2019**, *76*, 155–179. [CrossRef]
28. Shahsavari, D.; Shahsavari, M.; Li, L.; Karami, B. A Novel Quasi-3D Hyperbolic Theory for Free Vibration of FG Plates with Porosities Resting on Winkler/Pasternak/Kerr Foundation. *Aerosp. Sci. Technol.* **2018**, *72*, 134–149. [CrossRef]
29. Paik, J.K.; Seo, J.K. Nonlinear Finite Element Method Models for Ultimate Strength Analysis of Steel Stiffened-Plate Structures under Combined Biaxial Compression and Lateral Pressure Actions—Part II: Stiffened Panels. *Thin-Walled Struct.* **2009**, *47*, 998–1007. [CrossRef]
30. Putranto, T.; Kõrgesaar, M.; Jelovica, J.; Tabri, K.; Naar, H. Ultimate Strength Assessment of Stiffened Panel under Uni-Axial Compression with Non-Linear Equivalent Single Layer Approach. *Mar. Struct.* **2021**, *78*, 103004. [CrossRef]
31. Gavali, V.C.; Kubade, P.R.; Kulkarni, H.B. Mechanical and Thermo-Mechanical Properties of Carbon Fiber Reinforced Thermoplastic Composite Fabricated Using Fused Deposition Modeling Method. *Meter. Proc.* **2020**, *22*, 1786–1795. [CrossRef]
32. He, D.; Wang, Q.; Zhong, R.; Qin, B. A Unified Spectral-Geometric Model of FGM Double Conical/Cylindrical/Spherical Shell Coupled with Annular Plates. *Comput. Math. Appl.* **2023**, *143*, 348–371. [CrossRef]
33. He, D.; Wang, Q.; Zhong, R.; Qin, B. Vibration Analysis of Functionally Graded Material (FGM) Double Layered Floating Raft Structure by the Spectro-Geometric Method. *Structures* **2023**, *48*, 533–550. [CrossRef]

34. Zhang, S.; Bu, R.; Zhang, Z.; Gao, L.; Li, Z. A Systematic Model for the Mechanical Behavior of Thin-Walled Composite FGM Pipelines Subjected to Strike-Slip Faults in Geohazard Area. *Thin-Walled Struct.* **2024**, *202*, 112135. [CrossRef]
35. Penna, R.; Lovisi, G.; Feo, L. Dynamic Response of Multilayered Polymer Functionally Graded Carbon Nanotube Reinforced Composite (FG-CNTRC) Nano-Beams in Hygro-Thermal Environment. *Polymers* **2021**, *13*, 2340. [CrossRef]
36. Daikh, A.A.; Houari, M.S.A.; Karami, B.; Eltaher, M.A.; Dimitri, R.; Tornabene, F. Buckling Analysis of CNTRC Curved Sandwich Nanobeams in Thermal Environment. *Appl. Sci.* **2021**, *11*, 3250. [CrossRef]
37. Lovisi, G. Application of the Surface Stress-Driven Nonlocal Theory of Elasticity for the Study of the Bending Response of FG Cracked Nanobeams. *Compos. Struct.* **2023**, *324*, 117549. [CrossRef]
38. Miao, X.; Li, C.; Jiang, Y. Free Vibration Analysis of Metal-Ceramic Matrix Composite Laminated Cylindrical Shell Reinforced by CNTs. *Compos. Struct.* **2021**, *260*, 113262. [CrossRef]
39. Feng, Z.; Yang, X.; Keshavarzpour, H.; Ghasemi, A. Free Vibration Analysis of Hybrid CNT/GPL-Reinforced Porous Composite Plates under Fluid-Loading. *Aerosp. Sci. Technol.* **2024**, *148*, 109116. [CrossRef]
40. Wang, J.F.; Cao, S.H.; Zhang, W. Thermal Vibration and Buckling Analysis of Functionally Graded Carbon Nanotube Reinforced Composite Quadrilateral Plate. *Eur. J. Mech.-A/Solids* **2021**, *85*, 104105. [CrossRef]
41. Qin, B.; Zhong, R.; Wang, T.; Wang, Q.; Xu, Y.; Hu, Z. A Unified Fourier Series Solution for Vibration Analysis of FG-CNTRC Cylindrical, Conical Shells and Annular Plates with Arbitrary Boundary Conditions. *Compos. Struct.* **2020**, *232*, 111549. [CrossRef]
42. Zhong, R.; Wang, Q.; Tang, J.; Shuai, C.; Qin, B. Vibration Analysis of Functionally Graded Carbon Nanotube Reinforced Composites (FG-CNTRC) Circular, Annular and Sector Plates. *Compos. Struct.* **2018**, *194*, 49–67. [CrossRef]
43. Maoudj, S.; Tiberkak, R.; Lazar, M.E.; Ezzraimi, M.; Bachene, M.; Rechak, S. Free Vibration Analysis of Cracked Composite Plates Reinforced with CNTs Using Extended Finite Element Method (XFEM). *Mech. Adv. Mater. Struct.* **2023**. [CrossRef]
44. Ghadirian, H.; Mohebpour, S.; Malekzadeh, P.; Daneshmand, F. Nonlinear Free Vibrations and Stability Analysis of FG-CNTRC Pipes Conveying Fluid Based on Timoshenko Model. *Compos. Struct.* **2022**, *292*, 115637. [CrossRef]
45. TAYEB, T.S.; Amir, M.; Kim, S.-W.; Choi, D. An Analytical Modelling of Free Vibration in Porous FG-CNTRC Plate Resting on Elastic Foundations. *Mater. Today Commun.* **2024**, *40*, 109925. [CrossRef]
46. Cho, J.R. Nonlinear Bending Analysis of FG-CNTRC Plate Resting on Elastic Foundation by Natural Element Method. *Eng. Anal. Bound. Elem.* **2022**, *141*, 65–74. [CrossRef]
47. Duong, V.Q.; Tran, N.D.; Luat, D.T.; Thom, D.V. Static Analysis and Boundary Effect of FG-CNTRC Cylindrical Shells with Various Boundary Conditions Using Quasi-3D Shear and Normal Deformations Theory. *Structures* **2022**, *44*, 828–850. [CrossRef]
48. Duc, N.D.; Minh, P.P. Free Vibration Analysis of Cracked FG CNTRC Plates Using Phase Field Theory. *Aerosp. Sci. Technol.* **2021**, *112*, 106654. [CrossRef]
49. Taheri, M.H.; Memarzadeh, P. Effect of Crack on Shear Buckling of CNTRC Plates. *Int. J. Mech. Sci.* **2022**, *229*, 107519. [CrossRef]
50. Zeighami, V.; Jafari, M. Thermal Stress Analysis of Perforated Unsymmetric FG-CNTRC Plate Using a General Analytical Solution. *Thin-Walled Struct.* **2022**, *173*, 108956. [CrossRef]
51. Cong, P.H.; Trung, V.D.; Khoa, N.D.; Duc, N.D. Vibration and Nonlinear Dynamic Response of Temperature-Dependent FG-CNTRC Laminated Double Curved Shallow Shell with Positive and Negative Poisson's Ratio. *Thin-Walled Struct.* **2022**, *171*, 108713. [CrossRef]
52. Watts, G.; Kumar, R.; Singh, S.; Sengar, V.; Reddy, G.R.; Patel, S.N. Postbuckling and Postbuckled Vibration Behaviour of Imperfect Trapezoidal Sandwich Plates with FG-CNTRC Face Sheets under Nonuniform Loadings. *Aerosp. Sci. Technol.* **2022**, *127*, 107716. [CrossRef]
53. Lin, B.; Zhu, B.; Chen, B.; Han, J.; Li, Y. Nonlinear Primary Resonance Behaviors of Rotating FG-CNTRC Beams with Geometric Imperfections. *Aerosp. Sci. Technol.* **2022**, *121*, 107333. [CrossRef]
54. Liu, J.; Ke, L.-L.; Wang, Y.-S.; Yang, J.; Alam, F. Thermoelastic Frictional Contact of Functionally Graded Materials with Arbitrarily Varying Properties. *Int. J. Mech. Sci.* **2012**, *63*, 86–98. [CrossRef]
55. Wang, Z.-X.; Shen, H.-S. Nonlinear Vibration of Nanotube-Reinforced Composite Plates in Thermal Environments. *Comput. Mater. Sci.* **2011**, *50*, 2319–2330. [CrossRef]
56. Wang, Y.; Tham, L.; Cheung, Y. Beams and Plates on Elastic Foundations: A Review. *Prog. Struct. Eng. Mater.* **2005**, *7*, 174–182. [CrossRef]
57. Avramidis, I.E.; Morfidis, K. Bending of Beams on Three-Parameter Elastic Foundation. *Int. J. Solids Struct.* **2006**, *43*, 357–375. [CrossRef]
58. Karamanli, A.; Vo, T.P.; Eltaher, M.A. Comprehensive Analysis of Bio-Inspired Laminated Composites Plates Using a Quasi-3D Theory and Higher Order FE Models. *Thin-Walled Struct.* **2024**, *198*, 111735. [CrossRef]
59. Hasani Baferani, A.; Saidi, A.R.; Ehteshami, H. Accurate Solution for Free Vibration Analysis of Functionally Graded Thick Rectangular Plates Resting on Elastic Foundation. *Compos. Struct.* **2011**, *93*, 1842–1853. [CrossRef]
60. Selim, B.A.; Zhang, L.W.; Liew, K.M. Vibration Analysis of CNT Reinforced Functionally Graded Composite Plates in a Thermal Environment Based on Reddy's Higher-Order Shear Deformation Theory. *Compos. Struct.* **2016**, *156*, 276–290. [CrossRef]

61. Pasha Zanussi, V.; Shahverdi, H.; Khalafi, V.; Navardi, M.M. Nonlinear Flutter Analysis of Quadrilateral Plates Consisting of Functionally Graded Carbon Nanotubes Reinforced Composites Using Isogeometric Analysis. *Thin-Walled Struct.* **2024**, *198*, 111701. [CrossRef]
62. Lee, S.; Reddy, J. Vibration Suppression of Laminated Shell Structures Investigated Using Higher Order Shear Deformation Theory. *SMART Mater. Struct.* **2004**, *13*, 1176–1194. [CrossRef]

Journal of
*Marine Science and Engineering*

MDPI

*Article*

# A De-Nesting Hybrid Reliability Analysis Method and Its Application in Marine Structure

**Chenfeng Li [1,2], Tenglong Jin [1,2], Zequan Chen [1,2,*] and Guanchen Wei [1,2]**

[1] College of Shipbuilding Engineering, Harbin Engineering University, Harbin 150001, China; lichenfeng@hrbeu.edu.cn (C.L.); jine@hrbeu.edu.cn (T.J.); weiguanchen@hrbeu.edu.cn (G.W.)
[2] International Joint Laboratory of Architecture and Offshore Technology Between Harbin Engineering University and University of Lisbon, Harbin 150001, China
* Correspondence: chenzequan@hrbeu.edu.cn; Tel.: +86-182-4304-5947

**Abstract:** In recent years, marine structures have been widely used in the world, making significant contributions to the utilization of marine resources. In the design of marine structures, there is a hybrid reliability problem arising from aleatory uncertainty and epistemic uncertainty. In many cases, epistemic uncertainty is estimated by interval parameters. Traditional methods for hybrid reliability analysis usually require a nested optimization framework, which will lead to too many calls to the limit state function (LSF) and result in poor computational efficiency. In response to this problem, this paper proposes a de-nesting hybrid reliability analysis method creatively. Firstly, it uses the p-box model to describe the epistemic uncertainty variables, and then the linear approximation (LA) model and the two-point adaptive nonlinear approximation (TANA) model are combined to approximate the upper and lower bounds of LSF with epistemic uncertainty. Based on the first-order reliability method (FORM), an iterative operation is used to obtain the interval of the non-probability hybrid reliability index. The traditional nested optimization structure is effectively eliminated by the above approximation method, which efficiently reduces the times of LSF calls and increases the calculation speed while preserving sufficient accuracy. Finally, one numerical example and two engineering examples are provided to show the greater effectiveness of this method than the traditional nested optimization method.

**Keywords:** structural reliability; FORM; approximation model; de-nesting; marine structure

## 1. Introduction

In order to meet the needs of safety and economy of marine structures, uncertainty as a key factor in the design and manufacture of marine structures has been widely studied. Uncertainty in marine engineering includes aleatory uncertainty and epistemic uncertainty [1,2]. Aleatory uncertainty is caused by the inherent randomness of relevant variables in structures and ocean environments, which is usually described by an accurate probability distribution. Epistemic uncertainty is caused by the lack of sufficient knowledge or incomplete information about the variables, which can be described in the form of intervals and probabilistic methods [3].

In the past decades, aleatory uncertainty associated with marine structures has been a subject of extensive research. The traditional reliability assessment method is to use the FORM or the second-order reliability method (SORM) to calculate the structural reliability index after quantifying the uncertainty of random variables using probabilistic models [4], and the continuous development of reliability theory provides new tools for the reliability analysis of marine structures. Homaei and Najafzadeh [5] combined the artificial intelligence method and Monte Carlo sampling (MSC) to analyze the reliability of pile group scour depth being less than a limit value under regular waves. Aghatise Okoro et al. [6] used the polynomial chaos expansion (PCE) and Kriging method to construct a metamodel of the limit state function and reduced the computational time by the MSC method. Ming Cai Xu et al. [7] used the model correction factor method (MCFM) to constantly revise the results at the design points in the FORM iteration process to ensure that the results of the simplified model were close to the real model. The above research focuses on how to construct a surrogate model that is closer to the real model. However, their random variable distributions are usually based on empirical assumptions.

In fact, obtaining the accurate distribution of random variables requires a large number of sample data, which is difficult to achieve in practical engineering applications. In addition, it is often difficult to eliminate the uncertainty caused by structural manufacturing, installation errors and parameter measurement errors. To facilitate calculations, the aforementioned epistemic uncertainties are transformed into deterministic factors in many cases, leading to errors in the reliability assessment of marine structures. Debiao Meng et al. [8] analyzed the fatigue of offshore wind turbine structures and showed that a reliability assessment framework that considers both aleatory and epistemic uncertainties is more accurate and conservative. Therefore, it is necessary to incorporate the analysis method of epistemic uncertainty into the reliability assessment of marine structures.

In 1989, Ben-Haim and Elishakoff [9] proposed the idea of a non-probabilistic reliability model. Based on their study and the traditional probability theory, researchers began to propose a variety of non-probability models to describe the epistemic uncertain variables affecting structural reliability, such as interval theory [10,11], convex set theory [12,13], evidence theory [14,15] and p-box theory [16–18], etc. In many non-probabilistic models, it is simple and feasible to describe epistemic uncertain variables by using the interval model or the convex set model. The interval model and the convex set model envelope sample data points to obtain a hypercube or a super ellipsoid, which is the description of the epistemic uncertain variable. However, the interval model or the convex model only uses the edge data of the sample data, but does not make full use of information such as the dense distribution of internal data. This may lead to an overestimation of the epistemic uncertain variable and furthermore an excessively large interval of analysis results. The final result provides no effective guidance for solving practical engineering problems. Moreover, technicians can effectively estimate the distribution of epistemic uncertainty variables in some cases with their engineering experience and the analytical results of similar structures. However, this reference information cannot be effectively utilized with the interval model or the convex set model due to their theoretical limitations. As a new type of non-probability model theory, p-box theory describes the epistemic uncertain variable through the upper and lower probability boxes of the cumulative distribution function (CDF). Regardless of whether the distribution form is unknown or known, epistemic uncertain variables can be described by the p-box model [16]. In addition, p-box theory is compatible with other common non-probability theories. That is, other non-probability models can be transformed to the p-box model to some extent. Therefore, p-box theory can be used to establish a non-probability model with broader applicability for structural reliability analysis with epistemic uncertain variables. Research on the p-box model is burgeoning. Zhang et al. [19] obtained the statistical moments and the CDF of a response function with a non-parametric p-box variable based on the cumulative distribution function discretization (CDFD) method. Li and Jiang [20] combined the stochastic processes with p-box theory to propose a p-box-based imprecise stochastic process model, which is used for uncertainty analysis of structures under uncertain dynamic excitations or time-variant factors. Xiao et al. [21] proposed a collaborative interval quasi-Monte Carlo method (CIMCM) to deal with the reliability model with multiple types of epistemic uncertainty unified by p-boxes variables, combining Rosen's gradient projection method (RGPM) and collaborative optimization strategy, making the simulation convergence faster.

After the establishment of the non-probabilistic hybrid reliability model of the structure by using the traditional probabilistic model and the non-probability model, a hybrid structure reliability analysis is needed. At present, most of the literature needs to establish a two-layer nested optimization model [22–24]. The outer layer is the process of traditional structural reliability optimization, and the inner layer is the optimization model of the limit value of the LSF with the epistemic uncertain variables as the optimization vector. Due to the existence of the inner optimization structure, this nested optimization structure requires a large number of calls to the LSF, resulting in an inefficient calculation.

Referencing the FORM, this paper deals with the case of epistemic uncertainty variables with small uncertainty and combines the linear approximation (LA) model and the two-point adaptive nonlinear approximation (TANA) model to approximate the upper and lower limits of the response of the LSF. Consequently, a de-nesting analysis method of the non-probabilistic hybrid reliability index based on an approximation model is proposed. In this paper, the U space (standard normal distribution space) is converted back to original space of the LSF, and the boundary of the corresponding p-box variable is obtained. The approximation method eliminates the inner layer of the original nested optimization structure. The outer layer of the original nested optimization structure of the hybrid reliability analysis is transformed into a traditional optimization problem, with the constraint that the inner layer is expressed as an approximate formulation developed at the design point. The speed of non-probabilistic hybrid reliability analysis of structures will be effectively improved because of the elimination of nested structures. Finally, the proposed reliability assessment method is applied to a numerical example and two common marine structures to verify its effectiveness and superiority.

## 2. Non-Probabilistic Hybrid Structural Reliability

### 2.1. P-Box Theory

In p-box theory, the p-box model is a probability box enveloped by the upper and lower limits of the CDF, and a precise distribution form of variable distribution is difficult to obtain. When the distribution type of the variable is known, a more accurate p-box model can be obtained by estimating the interval of the distribution parameters of the distribution type. If the distribution type of the variable cannot be obtained, the Chebyshev inequality can be used to rigorously derive the upper and lower limits of the CDF of the variable, using information such as the origin moment of the sample data.

If the variable $y$ is described by the p-box model, it can be expressed as Equation (1) and Figure 1.

$$\underline{F_Y}(y) \leq F_Y(y) \leq \overline{F_Y}(y) \tag{1}$$

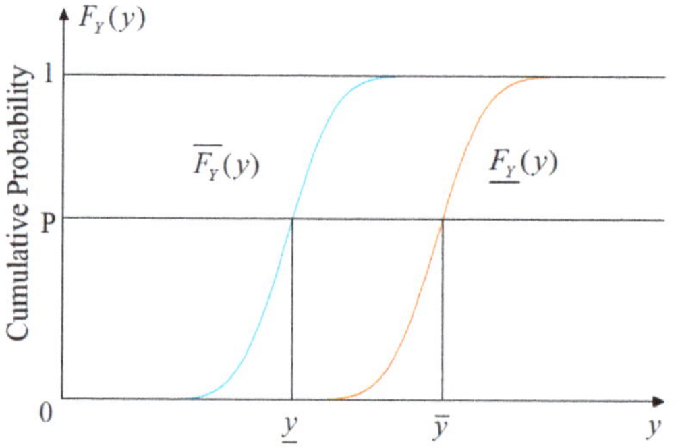


**Figure 1.** The upper and lower limits of the CDF of the p-box variable $y$.

The p-box model can be called a parametric p-box model when the distribution type of the epistemic uncertain variable is known and the distribution parameter is an estimation interval. When only the upper and lower limits of the CDF of the uncertain variable are known, the p-box model can be called a free p-box model.

The acquisition method of the free p-box model is generally Chebyshev's inequality, and the upper and lower limits of the CDF of the variable Y can be expressed as Equation (2a,b).

$$\underline{F_Y}(y) = \begin{cases} 0, & y < \mu + \sigma \\ 1 - \frac{\sigma^2}{(y-\mu)^2}, & y \geq \mu + \sigma \end{cases} \quad (2a)$$

as well as

$$\overline{F_Y}(y) = \begin{cases} \frac{\sigma^2}{(y-\mu)^2}, & y < \mu - \sigma \\ 1, & y \geq \mu - \sigma \end{cases} \quad (2b)$$

where $\overline{F_Y}(y)$ and $\underline{F_Y}(y)$ are the upper and lower limits of the CDF of the variable $y$, respectively, and $\mu$ and $\sigma$ are the mean and standard deviation of the epistemic uncertain variable $y$ in the sample data.

For the parametric p-box model, the CDF of the variable y can be expressed as Equation (3).

$$F_Y(y) = F_Y(y|\tau), s.t.\ \tau \in [\underline{\tau}, \overline{\tau}] \quad (3)$$

where $\tau$ is the distribution parameter of the distribution type that the variable $y$ follows. Due to the small amount of sample data and other possible reasons, this parameter is generally impossible to obtain accurately, but it can be estimated by the interval $[\underline{\tau}, \overline{\tau}]$.

### 2.2. Analytical Method

In the structural reliability analysis process, if the variable affecting the structural reliability contains $n$ random variables $X = (x_1, x_2, \cdots, x_n)$ (the exact CDF is known) and $m$ p-box variables $Y = (y_1, y_2, \cdots, y_m)$ (the exact CDF is unknown), the LSF $g(X, Y)$ of the structure can be obtained.

$$\begin{cases} g(X,Y) < 0, \cdots Structural \cdot failure \\ g(X,Y) = 0, \cdots\cdots\cdots Limit \cdot state \\ g(X,Y) > 0, \cdots Structural \cdot safety \end{cases} \quad (4)$$

To use FORM for structural reliability analysis, it is necessary to convert the LSF from the original space to the standard normal space (U space). The LSF in U space can be obtained by probability conversion, i.e., $G(U_X, U_Y) = g(X,Y)$ ($U_X$, $U_Y$) are independent standard normal distribution variables.

The independent variables $X$ and $Y$ after converting from the original space to the U space are given by

$$\begin{cases} U_X = \Phi^{-1}(F_X(X)) \\ U_Y = \Phi^{-1}(F_Y(Y)) \end{cases} \quad (5)$$

where $F_X(X)$ is the CDFs of the random variables $X$, $F_Y(Y)$ is the CDFs of the p-box variables $Y$ and $\Phi^{-1}$ is the inverse of the standard normal distribution.

According to Equation (5), there is,

$$\begin{cases} X = F_X^{-1}(\Phi(U_X)) \\ Y = F_Y^{-1}(\Phi(U_Y)) \in [Y^L, Y^R] = [\overline{F_Y}^{-1}(\Phi(U_Y)), \underline{F_Y}^{-1}(\Phi(U_Y))] \end{cases} \quad (6)$$

where $F_X^{-1}$ is the inverse functions of the CDFs of the random variables $X$, and $F_Y^{-1}$ is the inverse function of the CDFs of the p-box variables $Y$. $\overline{F_Y}^{-1}$ and $\underline{F_Y}^{-1}$ are, respectively, the inverse functions of the upper and lower limits of the CDFs of the p-box variables $Y$, and $Y^L$ and $Y^R$ are the lower and upper value limits of the p-box variables $Y$.

Further, the LSF for the structure can be expressed as Equation (7).

$$\begin{cases} g(X,Y) = G(U_X, U_Y) = G\left(\Phi^{-1}(F_X(X)), \Phi^{-1}(F_Y(Y))\right) \\ G(U_X, U_Y) = g(X,Y) = g\left(F_X^{-1}(f(U_X)), F_Y^{-1}(f(U_Y))\right) \end{cases} \quad (7)$$

Since the variables $Y \in [Y^L, Y^R]$ are p-box variables, the structural LSF can be written as Equation (8), and Figure 2 shows the surfaces of the non-probability hybrid LSF in the U space.

$$G(U_X, U_Y) = g(X,Y) \in [g^L, g^R] \tag{8}$$

where $g^L = \min_Y g(X,Y) = \min_G G(U_X, U_Y)$, and $g^R = \max_Y g(X,Y) = \max_G (U_X, U_Y)$.

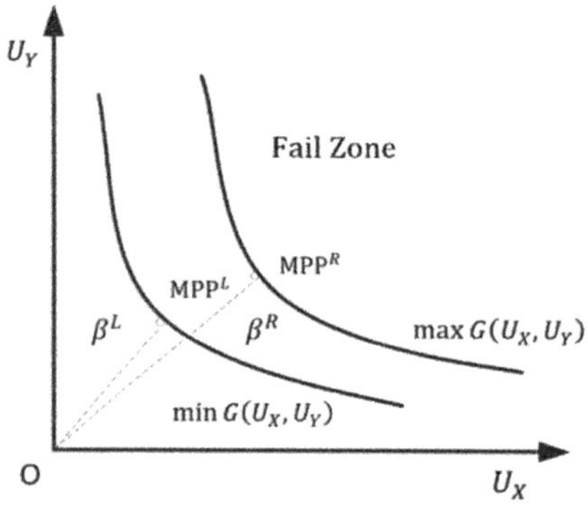


**Figure 2.** The surfaces of the non-probability hybrid LSF.

In Figure 2, MMP$^{\mathrm{L}}$ and MMP$^{\mathrm{R}}$ represent the maximum possible failure points (MPPs) of the lower LSF $g^L$ and upper LSF $g^R$, respectively. $\beta^L$ and $\beta^R$ are the upper and lower limits of the reliability index $\beta$. According to the FORM, the structural reliability index $\beta$ can be obtained, that is,

$$\begin{cases} \beta = \min_{U_X, U_Y} \sqrt{U_X^T U_X + U_Y^T U_Y} \\ s.t. G(U_X, U_Y) = g(X,Y) = 0 \end{cases} \tag{9}$$

It can be known from Equations (8) and (9) that the non-probability hybrid reliability index $\beta$ can be written as $\beta \in [\beta^L, \beta^R]$. Further, two nested optimization functions related to the upper and lower limits of the reliability index $\beta$ can be obtained, as shown in Equation (10a,b).

$$\begin{cases} \beta^L = \min\limits_{U_X, U_Y} \sqrt{U_X^T U_X + U_Y^T U_Y} \\ s.t. \quad G^L = g^L = \min\limits_Y g(X,Y) = 0 \end{cases} \tag{10a}$$

and

$$\begin{cases} \beta^R = \min\limits_{U_X, U_Y} \sqrt{U_X^T U_X + U_Y^T U_Y} \\ s.t. \quad G^R = g^R = \max\limits_Y g(X,Y) = 0 \end{cases} \tag{10b}$$

At the same time, it is necessary to discern the sign of the $\beta$. This requires obtaining the sign of the LSF at the origin of the standard normal space. According to Equation (11), the correct sign of $\beta$ can be determined.

$$\beta^{R(L)} = \mathrm{sgn}(G^{R(L)}(0,0))\beta^{R(L)} = \mathrm{sgn}(g^{R(L)}(\mu_X, \mu_Y))\beta^{R(L)} \tag{11}$$

In Equation (11), sgn represents a function that indicates the sign of $G^{R(L)}(0,0)$.

It can be known from Equation (10a,b) that the problem of hybrid structural reliability analysis with the p-box variables is a nested optimization problem. The inner layer is the process of solving the extremum of the LSF. For the above nested optimization problem, the computational workload of the traditional optimization algorithm is too heavy. In order to improve the calculation speed, the idea of an approximation model is used to approximate the upper and lower limits of the LSF. As such, the double-layer nested optimization structure is eliminated, the original optimization problem is transformed into a general optimization problem and finally a non-probabilistic hybrid reliability analysis method with p-box variables is constructed.

## 3. Approximation Model

In order to approximate the limit value of the LSF, it is necessary to use the approximation model. In this paper, the single-point approximation model and the two-point adaptive nonlinear approximation model are combined to approximate the LSF.

### 3.1. Single-Point Approximation Model

Based on the function value and gradient information at a single point, a single-point approximation model is established at the design point according to the first-order Taylor series. When using the traditional method optimization, since the search direction always needs to calculate the function value and the first derivative value in the optimization process, there is no need to add the calculation amount when constructing the single-point approximation model. There are a variety of single-point approximation models based on first-order Taylor expansion, including the linear approximation (LA) model, the reciprocal approximation model [25] and the conservative approximation model [26]. The single-point linear approximation model can be expressed as Equation (12).

$$g(X) = g(X_k) + \sum_{i}^{n} \frac{\partial g(X_k)}{\partial x_i}(x_i - x_{i,k}) \tag{12}$$

### 3.2. Two-Point Adaptive Nonlinear Approximation (TANA) Model

When using the single-point approximation model based on the first-order Taylor series, as the iteration proceeds, a new approximate model needs to be reconstructed at the new design point. The analysis information of the previous iteration step will be discarded and will not be used to improve the subsequent approximation model, resulting in relatively low fitting accuracy.

The two-point adaptive nonlinear approximation model is based on not only the information of the current design point, but also the information of the previous iteration point, and the nonlinear characteristic of the TANA model is automatically adjustable by the function value and the gradient values of the known point. Therefore, the TANA model is an adaptive approximation model whose simulation accuracy is higher than the single-point approximation model. The intermediate variable $z_i$ of the TANA model can be expressed as [27,28],

$$z_i = x_i^r, i = 1, 2, \cdots, n \tag{13}$$

For any variable $x_i$, the nonlinear index is equal to $r$.

When establishing a TANA model, the function values of the two points and the gradient values of the design points are necessary. First, the intermediate variable $z_i$ is used to represent the first-order Taylor expansion of the design point, and the variable $x_i$ is used in place of $z_i$ in the extended expression. Typically, one of the two points is selected as the new design point and the other point is selected as the design point of the previous iteration. Its mathematical expression is

$$g(X) = g(X_k) + \frac{1}{r}\sum_{i}^{n} x_{i,k}^{1-r} \frac{\partial g(X_k)}{\partial x_i}(x_i^r - x_{i,k}^r) \tag{14}$$

where $X_k$ denotes the design point of the $k$th iteration and $r$ denotes the nonlinear index. The nonlinear index will change during each iteration. The nonlinear index is determined by matching the function value at the previous design point $X_{k-1}$. That is, the difference between the approximation value and the exact value of $g(X_{k-1})$ is zero.

$$g(X_{k-1})-\left(g(X_k)+\frac{1}{r}\sum_{i}^{n}x_{i,k}^{1-r}\frac{\partial g(X_k)}{\partial x_i}(x_{i,k-1}^r-x_{i,k}^r)\right)=0 \tag{15}$$

In order to reduce the iterative calculation of the high-order polynomial approximation model, $r$ can be limited as $r\in[-5,5]$. This paper gives an idea of a solution. By constructing an optimization model, the nonlinear index $r$ is obtained.

$$\begin{cases} find\ r \\ \min\limits_{r}\left|g(X_{k-1})-\left(g(X_k)+\frac{1}{r}\sum\limits_{i}^{n}x_{i,k}^{1-r}\frac{\partial g(X_k)}{\partial x_i}(x_{i,k-1}^r-x_{i,k}^r)\right)\right| \\ s.t.-5\le r\le 5 \end{cases} \tag{16}$$

It can be understood from Equations (11) and (13) that the LA model is a special case of the TANA model. When the nonlinear index $r=1$, the TANA model is an LA model. The single-point approximation model is constructed according to the function value and the first-order derivative values at the design point, whereas in the TANA mode, the function value of another point is used to adjust the nonlinear index $r$ such that the approximation value is equal to the exact function value. Therefore, the approximation accuracy of TANA is higher than the single-point approximation model.

Figure 3 shows a comparison of the approximation of the function using the LA model and the TANA model at design point $x = 1$. Another point selected by the TANA model is $x_2=1.3$, and the nonlinear index is calculated as $r=-1.073$. It can be seen that the LA model can only approximate the original function with linear precision, and its approximation accuracy for strong nonlinear functions is poor. The approximation accuracy of the TANA model is significantly higher than the LA model, and there is no need to calculate the high-order derivative.

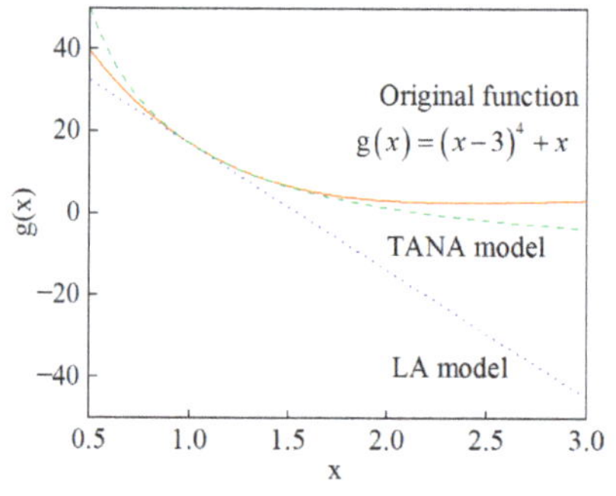


**Figure 3.** Comparison between LA model and TANA model at the design point $x = 1$.

## 4. De-Nesting Analysis Method

### 4.1. De-Nesting Theory

For the non-probabilistic hybrid structural reliability model, the LSF is $g(X,Y)$, and the method for calculating the structural reliability index is shown in Equation (10a,b). For $g(X,Y)$, since all of the epistemic uncertainty variables in practical engineering applications are of small uncertainty and the LA model is suitable for expanding the LSF in such scenarios, a satisfactory precision can be achieved. It is used to expand $g(X,Y)$ at $(X,Y^C)$, and its mathematical expression is

$$g(X,Y)\approx g(X,Y^c)+\sum_{j=1}^{m}\frac{\partial g(X,Y^c)}{\partial y_j}(y_j-y_j^c) \tag{17}$$

In Equation (17), there is $Y^c = (Y^L + Y^R)/2$ and $y_j^c = \left(y_j^L + y_j^R\right)/2$.

Using the TANA model to approximate $g(X, Y^C)$ at point $g(X_s, Y_s^C)$ to obtain Equation (18a), it is not difficult to substitute Equation (17) into Equation (18a), which combines the LA model and the TANA model to approximate the LSF, as shown in Equation (18b).

$$\begin{cases} g(X,Y^c) = g(X_s,Y_s^c) + R_n(X,r) + R_m(Y,r) \\ R_n(X,r) = \frac{1}{r}\sum_i^n x_{i,s}^{1-r}\frac{\partial g(X_s,Y_s^c)}{\partial x_i}(x_i^r - x_{i,s}^r) \\ R_m(Y^c,r) = \frac{1}{r}\sum_j^m (y_{j,s}^c)^{1-r}\frac{\partial g(X_s,Y_s^c)}{\partial \gamma_i^c}((y_j^c)^r - (y_{j,s}^c)^r) \end{cases} \tag{18a}$$

where s = 1, 2, 3 . . .

$$g(X,Y) \approx g(X_s,Y_s^c) + R_n(X,r) + R_m(Y,r) + \sum_{j=1}^m \frac{\partial g(X,Y^c)}{\partial y_j}(y_j - y_j^c) \tag{18b}$$

Substitute Equation (18b) into Equation (10a,b), and two de-nesting optimization functions related to the upper and lower limits of the reliability index can be obtained; for any nonlinear index $r = r_s$, there is

$$\begin{cases} \beta = \min\limits_{U_X,U_Y}\sqrt{U_X^T U_X + U_Y^T U_Y} \\ s.t. g^L \approx \min\limits_{Y}\left(g(X_s,Y_s^c) + R_n(X,r) + R_m(Y^c,r) + \sum\limits_{j=1}^m \frac{\partial g(X,Y^c)}{\partial y_j}(y_j - y_j^c)\right) = 0 \end{cases} \tag{19a}$$

$$\begin{cases} \beta = \min\limits_{U_X,U_Y}\sqrt{U_X^T U_X + U_Y^T U_Y} \\ s.t. g^R \approx \max\limits_{Y}\left(g(X_s,Y_s^c) + R_n(X,r) + R_m(Y^c,r) + \sum\limits_{j=1}^m \frac{\partial g(X,Y^c)}{\partial y_j}(y_j - y_j^c)\right) = 0 \end{cases} \tag{19b}$$

In Equation (19a,b), $X_S$ and $Y_S$ are obtained by probability conversion at the design expansion point $(U_{X,S}, U_{Y,S})$ after $s$th iterations. In other words, there is

$$X_s = F_X^{-1}(\Phi(U_{X,s})), Y_s \in [\overline{F_Y}^{-1}(\Phi(U_{Y,s})), \underline{F_Y}^{-1}(\Phi(U_{Y,s}))] \tag{20}$$

$Y_S^C$ is the median point of the p-box variable $Y$ after the $s$th iteration. There is

$$Y_s^c = (\overline{F_Y}^{-1}(\Phi(U_{Y,s}))) + \underline{F_Y}^{-1}(\Phi(U_{Y,s}))/2 \tag{21}$$

The nonlinear index $r$ can be solved with Equation (16).

$$\begin{cases} find\ r \\ \min\limits_{r}\left|g(X_{s-1},Y_{s-1}^c) - (g(X_s,Y_s^c) + R_n(X_{s-1},r) + R_m(Y_{s-1}^c,r))\right| \\ s.t. -5 \le r \le 5 \end{cases} \tag{22}$$

According to interval theory, the $g^L$ can be approximated using Equation (10a), which is

$$\begin{aligned} g^L &\approx \min_Y(g(X_s,Y_s^c) + R_n(X,r) + R_m(Y^c,r) + \sum_{j=1}^m \frac{\partial g(X,Y^c)}{\partial y_j}(y_j - y_j^c)) \\ &= g(X_S,Y_S^X) + R_n(X,r) + R_m(Y^c,r) - \sum_{j=1}^m\left|\frac{\partial g(X,Y^c)}{\partial y_j}\right|\left(\frac{y_j^R - y_j^L}{2}\right) \end{aligned} \tag{23a}$$

Similarly, $g^R$ should be

$$\begin{aligned} g^R &\approx \max_Y\left(g(X_s, Y_s^c) + R_n(X,r) + R_m(Y^c,r) + \sum_{j=1}^{m} \frac{\partial g(X,Y^c)}{\partial y_j}(y_j - y_j^c)\right) \\ &= g(X_S, Y_S^X) + R_n(X,r) + R_m(Y^c,r) + \sum_{j=1}^{m} \left|\frac{\partial g(X,Y^c)}{\partial y_j}\right| \left(\frac{y_j^R - y_j^L}{2}\right) \end{aligned} \tag{23b}$$

In the case where the form of the LSF is unknown or its first-order partial derivatives are difficult to obtain, the finite difference method can be used to address this problem.

$$\begin{cases} \frac{\partial g(X_s,Y_s^c)}{\partial x_i} = \frac{g(X_s+\Delta x_i,Y_s^c)-g(X_s,Y_s^c)}{\|\Delta x_i\|_2} \\ \frac{\partial g(X_s,Y_s^c)}{\partial y_j^c} = \frac{g\left(X_s,Y_s^c+\Delta y_j^c\right)-g(X_s,Y_s^c)}{\left\|\Delta y_j^c\right\|_2} \\ \frac{\partial g(X,Y^c)}{\partial y_j} = \frac{g\left(X,Y^c+\Delta y_j\right)-g(X,Y^c)}{\Delta y_{j2}} \end{cases} \tag{24}$$

where $\Delta x_i = (0,0,\cdots,\Delta_i,\cdots,0)$ and $\Delta y_j^c = \Delta y_j = (0,0,\cdots,\Delta_j,\cdots,0)$. $\Delta_i$ and $\Delta_j$ are constants that are small enough but not zero. $\|*\|_2$ expresses two-norm.

After going through the above process, for the double-layer nested optimization structure, the inner and outer optimization processes of the extreme value of LSF can be replaced with the approximation function, therefore eliminating the double-layer nested optimization structure and improving the calculation speed effectively. For the outer optimization problem, the TANA model is used to effectively avoid the problem that the convergence speed depends on the degree of approximation between the linearized LSF and the original nonlinear LSF. For each construction of the outer layer approximation model, $(n+m+1)$ times of solving the value of original LSF are needed at most. After the approximation model is constructed, for each time the extreme value of the approximation model is called, this process can be simplified to no more than $(m+1)$ times of original LSF value solution (one time of function value calculation at the median point and $m$ calls of function value due to the operation of the first-order partial derivative).

When using the present non-probabilistic hybrid reliability analysis methods, both the outer optimization process and the inner optimization processes require a large number of calls to the original LSF in order to obtain the limit of the LSF response. Compared with traditional nested analysis methods, the method proposed by this paper maintains the calculation accuracy and effectively reduces the number of calls to the LSF, which saves computing resources and increases computing speed.

### 4.2. De-Nesting Analysis Process

After using the LA model and the TANA model to approximate the LSF and obtain the limit value of the function response according to interval theory, the FORM can be used to obtain the structural reliability index $\beta$. According to References [25,29], the iterative equation of FORM can be expressed as

$$U_{k+1} = \frac{\nabla^T G(U_k) U_k - G(U_k)}{\nabla^T G(U_k) \alpha_k} \alpha_k \tag{25}$$

where $a_k$ is the negative unit normal vector at point $U_k$. According to the HL-RF method, $a_k$ can be expressed as the following equation.

$$\alpha_k = -\frac{\nabla G(U_k)}{\| \nabla G(U_k) \|} \tag{26}$$

Due to the use of TANA and LA models, there is a different $r_S$ in each approximation of the LSF. For any nonlinear index $r = r_S$, according to Equation (23a,b), the approximation expression of the upper and lower limits of the LSF can be known, that is, $G^R = g^R$ and $G^L = g^L$. The MMP can be calculated in combination with Equation (28). There are

$$U_{k+1}^{R(L)} = \frac{\nabla^T G^{R(L)}(U_k^{R(L)})U_k^{R(L)} - G^{R(L)}(U_k^{R(L)})}{\nabla^T G^{R(L)}(U_k^{R(L)})\alpha_k^{R(L)}}\alpha_k^{R(L)} \tag{27}$$

$$\alpha_k^{R(L)} = -\frac{\nabla G^{R(L)}(U_k^{R(L)})}{\nabla G^{R(L)}(U_k^{R(L)})} \tag{28}$$

where $U_k^{R(L)} = \left(U_{X,k}^{R(L)}, U_{Y,k}^{R(L)}\right)$. Similarly, the finite difference method can be used to obtain $\nabla G^{R(L)} = \left(\nabla G_1^{R(L)}, \nabla G_2^{R(L)}, \cdots \nabla G_p^{R(L)}, \cdots, \nabla G_{m+n}^{R(L)}\right)$

$$\nabla G_p^{R(L)}(U_{k,p}^{R(L)}) = \frac{G^{R(L)}(U_k^{R(L)} + \Delta U_{k,p}^{R(L)}) - G^{R(L)}(U_k^{R(L)})}{\Delta_p} \tag{29}$$

where $\Delta U_{k,p}^{R(L)} = (0, 0, \cdots, \Delta_p, \cdots, 0)$ and $\Delta_p$ is small enough but not zero.

For any nonlinear exponent $r = r_S$, the iterative convergence condition (1) can be set to

$$\begin{cases} \| U_{k+1}^{R(L)} - U_k^{R(L)} \|_2 < \delta \\ \beta_{k+1,s}^{R(L)} - \beta_{k,s}^{R(L)} < \varepsilon \end{cases} \tag{30}$$

$\beta_{k,s}^{R(L)}$ is expressed as the reliability index calculated by the $k$th iteration when the nonlinear index $r = r_S$. When the iteration meets the convergence condition (1), there should be $\beta_s^{R(L)} = \beta_{s,k}^{R(L)}$. At the same time, $U_s^{R(L)} = U_k^{R(L)}$ is used as the design expansion point to update the nonlinear index $r_s$ of the approximation model for the next iteration. The convergence condition (2) of the overall iterative algorithm can be set to

$$\beta_s^{R(L)} - \beta_{s-1}^{R(L)} < \varepsilon \tag{31}$$

The detailed process of the de-nesting analysis method is as follows, and the flow chart is as illustrated in Figure 4.

Step 1: Initialize parameters (1); let $S = 1$ record the number of iterations and provide an expansion point $U_S^{R(L)}$ and nonlinear index $r_S$ to provide the basis for calculation for the next step. In general, $U_1^{R(L)} = 0$ and $r_S = 1$, that is, linear expansion at the mean point of the variables.

Step 2: Variable conversion. Convert the vector $U_S^{R(L)}$ of the standard normal space back to the original space of the LSF to obtain $(X_S, X_S^C)$.

Step 3: At any point $(X, Y)$, using the LA model, the LSF approximation is expressed as an approximation expression for the p-box variables, i.e., Equation (17).

Step 4: According to the nonlinear index $r_S$, a TANA model for $g(X, Y^C)$ is established at the point $(X_S, X_S^C)$, i.e., Equation (18).

Step 5: Combine the LA model in the third step with the TANA model in the fourth step to approximate the extremum of the response of the LSF $g(X, Y)$ to any point $(X, Y)$.

Step 6: Initialize parameters (2), $k = 0, U_k^{R(L)} = U_s^{R(L)}$.

Step 7: According to Equations (27) and (28), the reliability index $\beta_{k,S}^{R(L)}$ can be iteratively calculated with the current nonlinear index $r_S$.

Step 8: Determine whether the convergence condition (1) is met. If the convergence condition (1) is met, the optimal reliability index $\beta_S^{R(L)}$ and the MPP $U_s^{R(L)}$ of the $r_S$ are obtained; go to the next step. If not, go back to Step 7 and re-iteratively perform the calculation.

Step 9: Determine whether the convergence condition (2) is met. If the convergence condition (2) is met, the final optimal non-probability hybrid reliability index $\beta^{R(L)}$ and the MPP are obtained; go to Step 11. If not, proceed to the next step.
Step 10: $S = S + 1$. Using the $U_S^{R(L)}$ calculated in Step 8 and the expansion point $U_{S-1}^{R(L)}$ of the previous iteration, update the nonlinear index $r_S$ by using Equation (20) to obtain the relevant parameters of the approximation model when proceeding to the next iteration. And return to Step 3 to start the calculation for the next iteration.
Step 11: Discriminate the sign of $\beta^{R(L)}$, following Equation (11), to obtain the final $\beta^{R(L)}$. The flow chart ends.

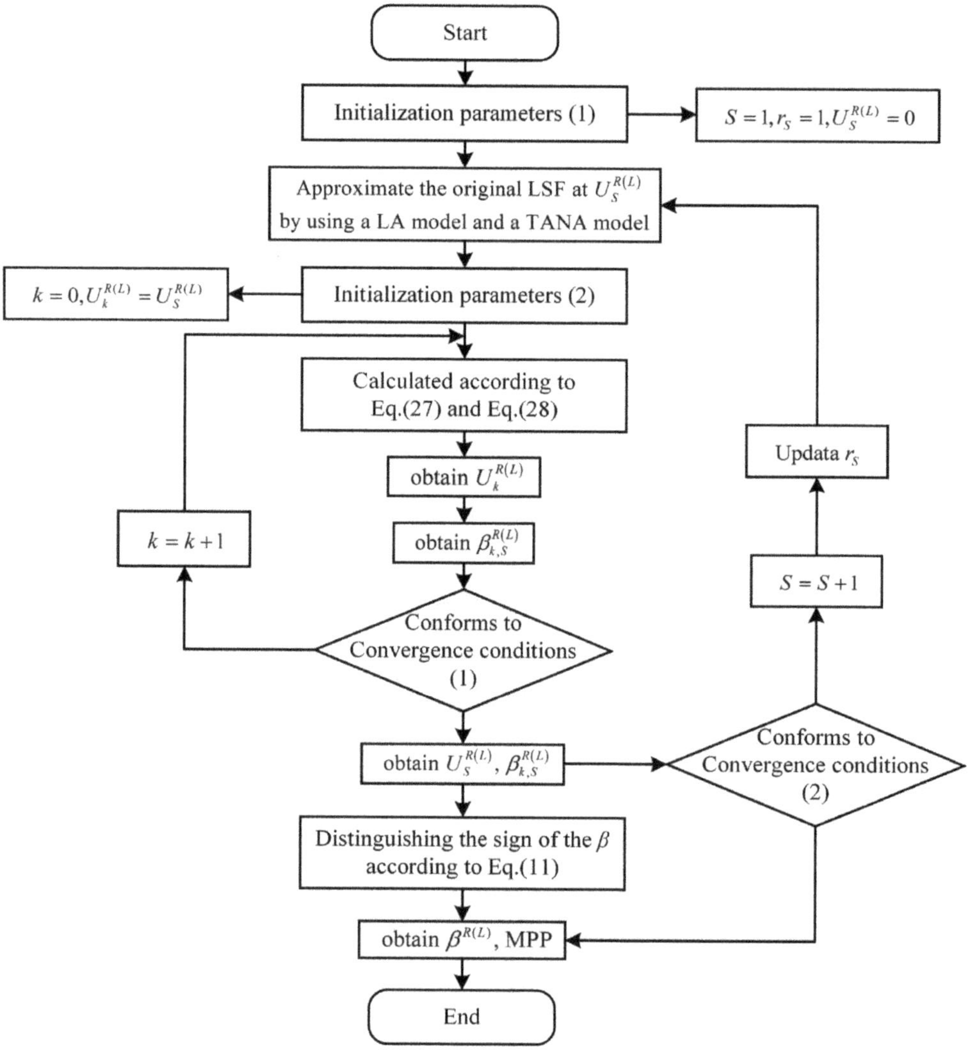


**Figure 4.** Flow chart of the de-nesting analysis method of the non-probability hybrid reliability index.

## 5. Numerical Examples and Discussion

### 5.1. Numerical Examples

The LSF has been known as the following equation.

$$g(X,Y) = x_1 - x_2^2 + x_3^3 + y_1^2 - y_2^2 \quad (32)$$

where $X = (x_1, x_2, x_3)$ is a random variables vector with known exact distributions, and $Y = (y_1, y_2)$ is a non-probabilistic p-box variables vector with unknown exact distributions. More specific information about the variables is shown in Table 1.

**Table 1.** Distribution of variables in numerical examples.

| Variables | Mean $\mu$ | Standard Deviation $\sigma$ | Distribution |
|---|---|---|---|
| $x_1$ | 2.0 | 0.05 | Normal |
| $x_2$ | 1.5 | 0.05 | Normal |
| $x_3$ | 1.2 | 0.05 | Normal |
| $y_1$ | [1.2, 1.4] | 0.05 | Normal |
| $y_2$ | [1.4, 1.6] | 0.05 | Normal |

The bounds on CDF of two p-box variables $y_1$ and $y_2$ are shown in Figure 5.

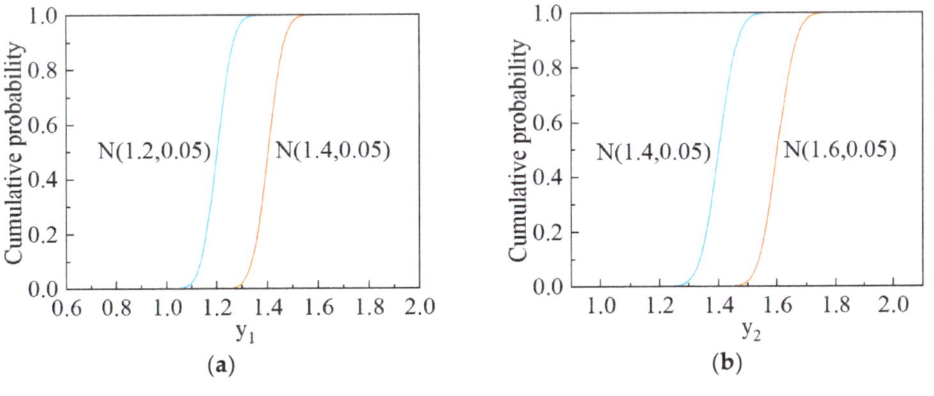


**Figure 5.** Upper and lower limits of the CDF of the p-box variable in numerical examples. (**a**) Normal distribution variable $y_1$; (**b**) normal distribution variable $y_2$.

In the convergence conditions (1) and (2), there are $\delta = 10^{-3}$, $\varepsilon = 10^{-3}$. The structural hybrid reliability index $\beta^{R(L)}$ can be calculated as $\beta \in [\beta^L, \beta^R] = [1.0811, 4.5954]$, according to the de-nesting analysis method of the non-probability hybrid reliability index in this paper. In contrast, the traditional nested method is also used, in which the precise limit value of the LSF is used. After combining it with the FORM of [30], there is $\beta \in [\beta^L, \beta^R] = [1.0812, 4.5934]$.

In the process of obtaining the lower limit of the reliability index $\beta^L$, the nonlinear index $r$ of the proposed method is [1, 3.60, 1] in three iterations. In the process of obtaining $\beta^R$, the nonlinear index $r$ is $[1, 3.27, -1, 1]$ in four iterations. According to the comparison in Figure 6, it can be known that the convergence speed of the calculation is not limited to the linear approximation accuracy of the traditional FORM, and the convergence speed compared with the traditional FORM is greatly improved, as the TANA model is used to construct the approximation function. At the same time, since the extreme value of the LSF is approximated by the approximation models, there is no need to perform an interval analysis of the extreme value of the LSF. The main workload of each iteration step is only a few calculations of the LSF, therefore effectively reducing the resources and time required for calculations. More details of the two methods are in Tables 2 and 3.

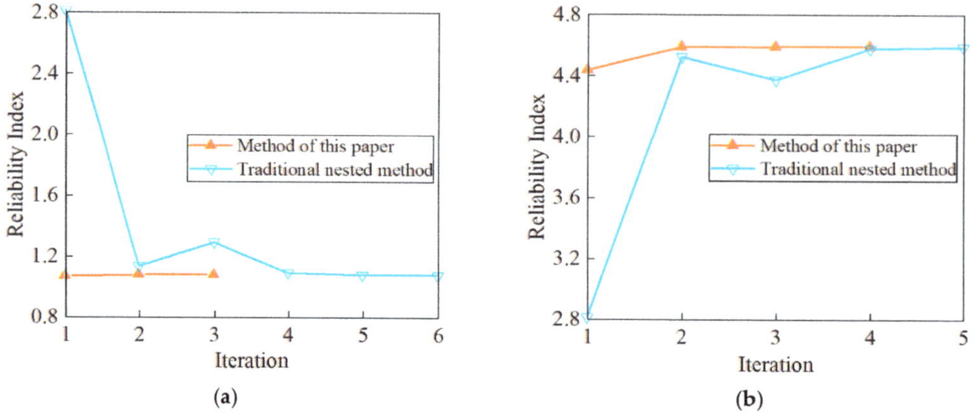


**Figure 6.** Comparison of the method of this paper with the traditional nested method when calculating the (**a**) min reliability index $\beta^L$ and the (**b**) max reliability index $\beta^R$.

**Table 2.** Comparison of the method of this paper with the traditional nested method when solving the min reliability index $\beta^L$.

| Method | $\beta^L$ | Iterations | MPP $(U_X, U_Y)$ | Number of Function Calls |
|---|---|---|---|---|
| Method of this paper | 1.0811 | 3 | [−0.1646, 0.5022, −0.6720, −0.3887, 0.5357] | 134 |
| Traditional nested method | 1.0812 | 6 | [−0.1646, 0.5022, −0.6720, −0.3887, 0.5357] | 468 |

**Table 3.** Comparison of the method of this paper with the traditional nested method when solving the max reliability index $\beta^R$.

| Method | $\beta^R$ | Iterations | MPP $(U_X, U_Y)$ | Number of Function Calls |
|---|---|---|---|---|
| Method of this paper | 4.5954 | 4 | [−0.7325, 2.3711, −2.5317, −1.9109, 2.2130] | 247 |
| Traditional nested method | 4.5934 | 5 | [−0.7321, 2.3699, −2.5309, −1.9100, 2.2121] | 612 |

In order to express the accuracy of the proposed method for dealing with epistemic uncertainty variables with small uncertainty more intuitively, we use the above numerical example and compare the method proposed in this paper with traditional nested methods in adjusting the uncertainty of epistemic uncertainty variables.

Uncertainty is defined as Equation (33) and the details are shown in Table 4.

$$\text{Uncertainty} = \frac{2(\mu^R - \mu^L)}{\mu^R + \mu^L} \times 100\% \tag{33}$$

**Table 4.** Distribution of epistemic uncertain variables under different uncertainties.

| Uncertainty | Variable | Mean $\mu$ | Variable | Mean $\mu$ |
|---|---|---|---|---|
| 0% | | 1.300 | | 1.500 |
| 5% | | [1.235, 1.365] | | [1.425, 1.575] |
| 10% | $y_1$ | [1.170, 1.430] | $y_2$ | [1.350, 1.650] |
| 15% | | [1.105, 1.495] | | [1.275, 1.725] |
| 20% | | [1.040, 1.560] | | [1.200, 1.800] |

According to Figure 7, the interval of the non-probabilistic hybrid reliability index gradually increases with the uncertainties of the epistemic uncertain variables. In addition, under different uncertainties of epistemic uncertainty variables, the method proposed in this paper can maintain high calculation accuracy in comparison to traditional nested methods by effectively eliminating nested analysis structures.

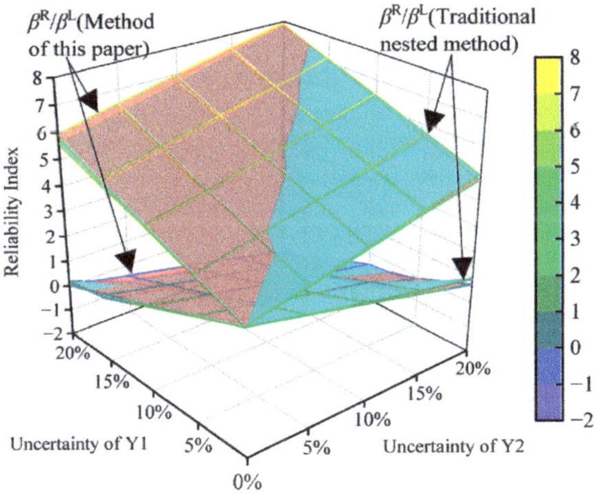


**Figure 7.** Upper and lower limits of the reliability index for epistemic uncertain variables with different degrees of uncertainty (method of this paper and the traditional nested method).

Furthermore, when the uncertainty of the epistemic uncertainty variable is too large, the approximation accuracy of the LA model is insufficient. The interval subdivision strategy can be used to subdivide the epistemic uncertainty variable $y_j$ into $a_j$ subintervals. It can be deduced that the nested structure of the traditional method can be eliminated by solving the LSF value $\prod_{j=1}^{m} a_j \times (m+1)$ times. Even if the calls of LSF are increased due to the interval subdivision strategy, the method of this paper still greatly improves the analysis speed compared with the traditional nested method.

### *5.2. Application to the Ultimate Strength of Stiffened Panels*

The stiffened panel subjected to longitudinal compression in the literature [31] is used as the reliability analysis object to verify the method proposed in this paper, as shown in Figure 8. The relevant parameters of stiffened panel are shown in Table 5. Among them, $a$, $b$, $h_w$ and $b_f$ are constants, representing the plate length, plate width, plate thickness, web height and flange width, respectively; $\sigma_{Yeq}$ and $E$ are random variables that denote the equivalent yield strength and elastic modulus of material, respectively; $t_p$, $t_w$ and $t_f$ are p-box variables with unknown precise distribution, which denote the thickness of panel, web and flange, respectively.

**Table 5.** Distribution of constants and variables in numerical examples.

| **Constants** | **Value** | **Unit** | - | |
|---|---|---|---|---|
| $a$ | 4300 | mm | | |
| $b$ | 815 | mm | | |
| $h_w$ | 463 | mm | | |
| $b_f$ | 172 | mm | | |
| **Variables** | **Mean $\mu$** | **Coefficient of variation (CV)** | **Distribution** | **Unit** |
| $\sigma_{Yeq}$ | 315 | 0.05 | Normal | MPa |
| $E$ | 205,800 | 0.05 | Normal | MPa |
| $t_p$ | [17.622, 17.978] | 0.10 | Normal | mm |
| $t_w$ | [7.920, 8.080] | 0.05 | Normal | mm |
| $t_f$ | [16.830, 17.170] | 0.05 | Normal | mm |

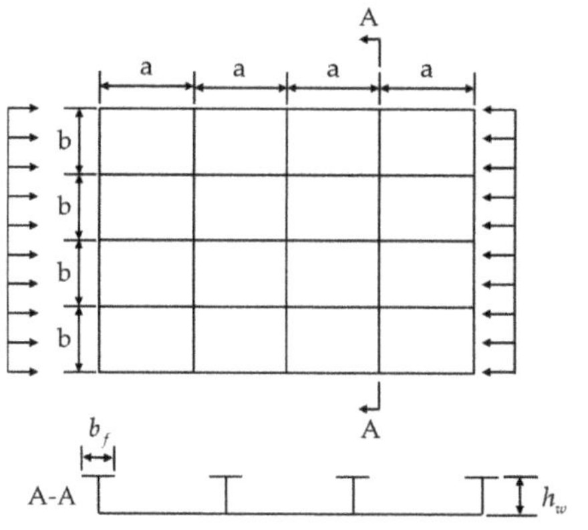


**Figure 8.** Schematic view of stiffened panel subjected to longitudinal compression and its geometric parameters.

In [32], the empirical formula for the prediction of ultimate strength of stiffened panels is obtained by using the finite element simulation method. It is assumed that the ratio of the plate strength to the yield strength cannot be less than 0.7, and the LSF can be expressed as follows:

$$g = \sigma_{xu}/\sigma_{Yeq} - 0.7 = \frac{1}{0.8884 + e^{\lambda^2}} + \frac{1}{0.4121 + e^{\sqrt{\beta}}} - 0.7 \tag{34}$$

where $\sigma_{xu}$ is the ultimate strength of stiffened panel subjected to axial compression, and $\beta$ and $\lambda$ are the plate slenderness ratio and stiffener slenderness ratio, respectively.

$$\begin{cases} \beta = (b/t_p)\sqrt{\sigma_{Yeq}/E} \\ \lambda = (a/\pi r)\sqrt{\sigma_{Yeq}/E} \end{cases} \tag{35}$$

where $r = \sqrt{I/A}$ is the radius of gyration of the stiffened panel, $I$ is the moment of inertia and $A$ is the cross-sectional area.

In the convergence conditions (1) and (2), there are $\delta = 10^{-3}$, $\varepsilon = 10^{-3}$. The hybrid reliability index is calculated using the method proposed in this paper. After three iterations, there is $\beta \in [\beta^L, \beta^R] = [2.4519, 2.6501]$. When is $\beta^R$ solved, the nonlinear index $r$ is [0.119, 0.233, 1.000] in three iterations. When solving $\beta^L$, the nonlinear index $r$ is [0.107, 0.204, 1.000]. Using the traditional nested method, after four iterations, $\beta \in [\beta^L, \beta^R] = [1.0453, 4.8904]$ is obtained. According to Tables 6 and 7, the method calls the LSF 223 times and 223 times to solve the lower and upper limits of the reliability index, respectively, while the traditional nested method requires 360 and 288 calls of the LSF.

**Table 6.** Comparison of the method of this paper with the traditional nested method when solving the min reliability index $\beta^L$.

| Method | $\beta^L$ | Iterations | MPP $(U_X, U_Y)$ | Number of Function Calls |
|---|---|---|---|---|
| Method of this paper | 2.4519 | 3 | [0.5959, −0.6337, −2.2916, 0.0059, −0.0635] | 223 |
| Traditional nested method | 2.4544 | 20 | [0.5961, −0.6337, −2.2941, 0.0060, −0.0635] | 360 |

**Table 7.** Comparison of the method of this paper with the traditional nested method when solving the max reliability index $\beta^R$.

| Method | $\beta^R$ | Iterations | MPP $(U_X, U_Y)$ | Number of Function Calls |
|---|---|---|---|---|
| Method of this paper | 2.6501 | 3 | [0.6431, −0.6876, −2.4763, 0.0067, −0.0677] | 223 |
| Traditional nested method | 2.6436 | 16 | [0.6425, −0.6864, 2.4698, 0.0066, −0.0681] | 288 |

### 5.3. Application to the Ultimate Strength of Ship Hull Girder

The ultimate strength of a ship hull girder is an important attribute to ensure navigation safety. Therefore, the proposed method is applied to the reliability analysis of ship hull girders with external loads arising from hydrostatic pressures and wave-induced forces. The external load caused by hydrostatic pressure is called the still water bending moment (SWBM), while the external load caused by waves is usually considered only as the vertical wave-induced bending moment (VWBM), because the effect of horizontal bending and shear on the hull beam is generally considered to be negligible [33]. The LSF of a hull girder's ultimate strength can be represented by the following equation [34].

$$g = \chi_u M_u - (\chi_s M_s + \chi_w \varphi_w M_w) \tag{36}$$

where $M_u$ is the total longitudinal ultimate bending moment of the hull girder; $M_s$ and $M_w$ are the SWBM and VWBM, respectively, whose distribution parameters can be calculated using Equations (A2)–(A10) in Appendix A; $\chi_u$, $\chi_s$ and $\chi_w$ are model uncertainty factors corresponding to $M_u$, $M_s$ and $M_w$, respectively, which are generally regarded as normally distributed variables; $\varphi_w$ is the load reduction factor caused by the superposition of SWBM and VWBM, which is set to 0.9 in this paper

The ship hull girder calculated in this paper is shown in Figure 9a, and more details about its structure can be obtained from Appendix A. The direct calculation method is used to solve $M_u$ in Equation (36), which obtains the maximum moment capacity of ship hull girders by assuming the cross-sectional stress distribution of hull girder in the ultimate state [35], as shown in Figure 9b. In this paper, the hull girder in the sagging limit state is considered, where both the deck and the adjacent side have buckled while the outer bottom has yielded, and the height range of D/2 near the neutral axis remains in the elastic region.

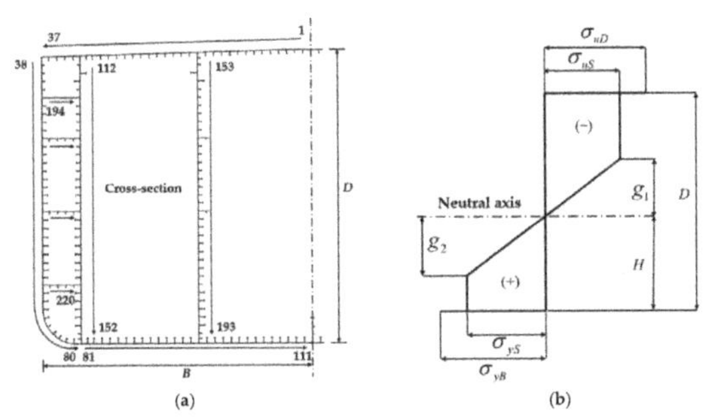


**Figure 9.** (**a**) Schematic view of the cross-section of a ship hull girder; (**b**) cross-sectional stress distribution of the ship hull girder in the sagging limit state, as assumed in this paper.

The equation for calculating the hull girder's ultimate moment $M_u$ is as follows:

$$M_u = \sum_i \sigma_{u_i}^c A_{ps_i} z_i + \sum_j \sigma_{u_j}^t A_{ps_j} z_j + \sum_k \sigma^e A_{ps_k} z_k \tag{37}$$

where $A_{ps_i}$ is the area of the stiffener plate or hard Angle element, $z$ represents the vertical distance from the structural element's height to the neutral axis, $i$, $j$ and $k$ are the compressive, tensile and elastic zones, respectively. $\sigma_u^c$ represents the ultimate buckling strength in the compressive zone, calculable using Equation (A1) in Appendix A; $\sigma_u^t$ denotes the yield strength in the tensile zone and $\sigma^e$ represent elastic stress distribution in the elastic zone.

Table 8 shows the main particulars of ship in the case calculated in this paper and Table 9 lists all relevant parameters considered in the reliability analysis. Among them, model uncertainty factors $\chi_u$, $\chi_s$, $\chi_w$, SWBM $M_S$, VWBM $M_w$, elastic modulus $E$ and yield strength $\sigma_{Yeq}$ are random variables; panel thickness $t_p$, web thickness $t_w$ and flange thickness $t_f$ are p-box variables with unknown precise distribution. $t_p^n$, $t_w^n$ and $t_f^n$ present the design thickness of every panel, web and flange, respectively. The distribution parameters of the material properties of the hull structure and the plate thickness are selected by referring to a previous study [33].

**Table 8.** Distribution of variables in numerical examples.

| Description | Symbol | Value | Unit |
|---|---|---|---|
| Length between perpendiculars | $L_{PP}$ | 328.2 | m |
| Molded breadth | $B$ | 62 | m |
| Molded depth | $D$ | 32.8 | m |
| Scantling draught | $T_{SC}$ | 24.847 | m |
| Block coefficient | $C_B$ | 0.9148 | - |
| Frame spacing | $L_{Fs}$ | 5.65 | m |

**Table 9.** Distribution of variables in numerical examples.

| Variables | Mean $\mu$ | CV | Distribution | Unit |
|---|---|---|---|---|
| $\chi_u$ | 1 | 0.05 | Normal | - |
| $\chi_s$ | 1 | 0.10 | Normal | - |
| $\chi_w$ | 1 | 0.24 | Normal | - |
| $M_S$ | 7.33 | 0.13 | Gumbel | $\times 10^3$ MN·m |
| $M_w$ | 1.11 | 0.05 | Gumbel | $\times 10^4$ MN·m |
| $E$ | 2.06 | 0.03 | Lognormal | $\times 10^5$ MPa |
| $\sigma_{Yeq}$ | 235 | 0.05 | Lognormal | MPa |
| | 350 | 0.05 | Lognormal | MPa |
| $t_p$ | $[0.95, 1.05] \times t_p^n$ | 0.05 | Normal | mm |
| $t_w$ | $[0.95, 1.05] \times t_w^n$ | 0.05 | Normal | mm |
| $t_f$ | $[0.95, 1.05] \times t_f^n$ | 0.05 | Normal | mm |

In the convergence conditions (1) and (2), there are $\delta = 10^{-3}$, $\varepsilon = 10^{-3}$. The hybrid reliability index is calculated by using the method proposed in this paper. After five iterations, there is $\beta \in [\beta^L, \beta^R] = [3.7308, 4.4726]$. When $\beta^R$ is solved, the nonlinear index $r$ is [1.339, 0.363, −0.003, 0.067, 1.463] in five iterations. When solving $\beta^L$, the nonlinear index $r$ is [1.290, 0.233, −0.008, 0.059, 1.612]. This proves that the limit response surface of this case has strong nonlinear characteristics. Using the traditional nested method, after 28 and 24 iterations, $\beta \in [\beta^L, \beta^R] = [3.6839, 4.4122]$ is obtained. According to Tables 10 and 11, the method calls the LSF 1942 times and 2418 times to solve the lower and upper limits of the reliability index, respectively, while the traditional nested method requires 5532 and 5676 calls of the LSF. It can be found that when solving the reliability index of complex implicit LSFs, the accuracy of the proposed method is not much different from that of the traditional method, but it has a very obvious effect on reducing the number of function calls.

**Table 10.** Comparison of the method of this paper with the traditional nested method when solving the min reliability index $\beta^L$.

| Method | $\beta^L$ | Iterations | MPP $U_Y$ | Number of Function Calls |
|---|---|---|---|---|
| Method of this paper | 3.7308 | 5 | [−1.0579, −0.3323, −0.1092] | 1942 |
| Traditional nested method | 3.6839 | 28 | [−1.0321, −0.3239, −0.1065] | 5532 |

**Table 11.** Comparison of the method of this paper with the traditional nested method when solving the max reliability index $\beta^R$.

| Method | $\beta^R$ | Iterations | MPP $U_Y$ | Number of Function Calls |
|---|---|---|---|---|
| Method of this paper | 4.4726 | 5 | [−1.0660, −0.3479, −0.1143] | 2418 |
| Traditional nested method | 4.4122 | 24 | [−1.0443, −0.3401, −0.1119] | 5676 |

## 6. Conclusions

In this paper, a de-nesting analysis method of the non-probabilistic hybrid reliability index based on an approximate model is proposed for the reliability assessment of marine structures. The LA model and the TANA model are combined to approximate the extreme response value of the LSF quickly, in which it is only necessary to calculate the first-order partial derivative value of the LSF for a few times during each iteration calculation. As supported by the numerical examples and two engineering examples, as well as the comparison with the traditional nested structural reliability analysis methods, we show that the proposed method will reduce the number of LSF calls effectively and save computing resources. In cases where the LSF is an implicit structure and finite element analysis is required, an efficient reduction in the number of LSF calls means effectively reducing the number of finite element analyses and improving calculation efficiency compared to the traditional method.

In addition, the proposed method eliminates the inner structure of the traditional reliability analysis by using an LA model. Therefore, it is generally applicable to the hybrid structural reliability analysis with small uncertainty. For general marine structure problems, small uncertainty is very easy to implement, because excessive uncertainty leads to an excessively large interval of the structural reliability index, which is pointless for engineering guidance. Combined with the engineering example, the proposed algorithm has good convergence and calculation accuracy.

One limitation of the paper is that the adopted approximation method does not consider the dependence of variables when approximating the LSF extreme value by the LA model. The approximation accuracy decreases when there is strong dependence between variables. Thus, a de-nesting hybrid reliability analysis that considers strong variable dependence will be conducted in the future.

**Author Contributions:** Conceptualization, C.L. and Z.C.; methodology, Z.C.; software, T.J.; validation, T.J. and G.W.; formal analysis, C.L. and Z.C.; investigation, G.W.; resources, Z.C.; data curation, T.J. and Z.C.; writing—original draft preparation, T.J.; writing—review and editing, C.L. and T.J.; visualization, T.J. and G.W.; supervision, Z.C.; project administration, Z.C. All authors have read and agreed to the published version of the manuscript.

**Funding:** This work was supported by The development and application project of ship CAE software (CBZ01N23-01), and National Key R&D Program of China (2022YFB3306200), and Fundamental Research Funds for the Central Universities (3072024CFJ0111), and National Natural Science Foundation of China (Grant No. 52171305).

**Institutional Review Board Statement:** Not applicable.

**Informed Consent Statement:** Not applicable.

**Data Availability Statement:** Relevant research data are mostly displayed in this paper. If you require other data, please contact the corresponding author by email.

**Conflicts of Interest:** The authors declare no conflicts of interest.

## Appendix A

This appendix describes the information related to the calculation of the ultimate strength of the hull girder and the calculation method of the distribution parameters of the SWBM and VWBM.

*Appendix A.1. Utimate Strength of Hull Girder*

The ultimate strength of structure elements can be obtained using the following empirical Equation [36]:

$$\sigma_u^c/\sigma_{Yeq} = \frac{1}{\sqrt{0.995 + 0.936\lambda^2 + 0.170\beta^2 + 0.188\lambda^2\beta^2 - 0.067\lambda^4}} \quad \text{(A1)}$$

Tables A1 and A2 report the design values of the geometric properties of the structural members of the ship hull girder calculated in this paper.

**Table A1.** The design values of the geometric properties of the structural members in Figure 9a—Part 1.

| Element Member | Plating | | Stiffener | | | |
|---|---|---|---|---|---|---|
| | $b_{p,n}$ (mm) | $t_{p,n}$ (mm) | $h_{w,n}$ (mm) | $t_{w,n}$ (mm) | $b_{f,n}$ (mm) | $t_{f,n}$ (mm) |
| 1–3, 7–10, 12–21, 23–28 | 900 | 24 | 450 | 30 | - | - |
| 4–6, 11, 22 | 900 | 24 | 450 | 30 | 150 | 30 |
| 15–16, 31 | 450 | 24 | - | - | - | - |
| 29 | 800 | 24 | 450 | 30 | - | - |
| 30 | 700 | 24 | 450 | 30 | - | - |
| 32, 134, 145–146 | 450 | 23 | - | - | - | - |
| 33 | 900 | 23 | 450 | 20 | 150 | 30 |
| 34–35 | 900 | 23 | 450 | 30 | - | - |
| 36 | 850 | 23 | 450 | 30 | - | - |
| 37 | 400 | 23 | - | - | - | - |
| 38, 43–44, 49–50, 59, 194, 196 | 465 | 30 | - | - | - | - |
| 39 | 930 | 30 | 281 | 14 | 120 | 19 |
| 40–42 | 930 | 30 | 328 | 14 | 130 | 22 |
| 45–48, 51–52 | 930 | 30 | 470 | 14 | 140 | 25 |
| 53–54 | 930 | 30 | 490 | 14 | 140 | 25 |
| 55–56 | 930 | 30 | 550 | 14 | 140 | 25 |
| 57–58 | 930 | 30 | 590 | 14 | 140 | 25 |
| 60 | 490 | 30 | - | - | - | - |
| 61–63, 65 | 980 | 30 | 632 | 18 | 150 | 28 |
| 64 | 980 | 30 | 630 | 18 | 185 | 30 |
| 66 | 905 | 30 | 630 | 18 | 185 | 30 |
| 67 | 830 | 30 | 638 | 18 | 185 | 22 |
| 68 | 830 | 30 | 570 | 14 | 140 | 25 |
| 69–70 | 220 | 30 | - | - | - | - |
| 71 | 930 | 30 | 632 | 18 | 185 | 28 |
| 72 | 930 | 30 | 630 | 30 | 185 | 30 |
| 73 | 1041 | 23.5 | 632 | 18 | 185 | 28 |
| 74 | 1041 | 23.5 | 576 | 18 | 185 | 24 |
| 75 | 941 | 23.5 | 410 | 18 | 150 | 20 |
| 76 | 831 | 23.5 | 410 | 18 | 150 | 20 |
| 77 | 892 | 23.5 | 410 | 18 | 150 | 20 |
| 78 | 784 | 23.5 | 410 | 18 | 150 | 20 |
| 79 | 1041 | 23.5 | 578 | 18 | 180 | 22 |
| 80–81 | 400 | 23.5 | - | - | - | - |
| 82–83 | 750 | 23.5 | 668 | 20 | 185 | 32 |
| 84–104 | 900 | 23.5 | 668 | 20 | 185 | 32 |
| 96, 193 | 260 | 24 | - | - | - | - |
| 97 | 225 | 23.5 | 668 | 20 | 185 | 32 |
| 105–110, 186–190 | 900 | 26.5 | 668 | 20 | 185 | 32 |
| 111–112 | 485 | 24 | - | - | - | - |
| 113–116 | 930 | 21 | 500 | 15 | 160 | 20 |
| 117–118, 123–124, 133,164–165, 174–175, 184–185, 215 | 465 | 21 | - | - | - | - |
| 119–122 | 930 | 21 | 415 | 14.5 | 135 | 24.5 |
| 125–132 | 930 | 21 | 500 | 14 | 140 | 25 |

**Table A2.** The design values of the geometric properties of the structural members in Figure 9a—Part 2.

| Element Member | Plating | | Stiffener | | | |
|---|---|---|---|---|---|---|
| | $b_{p,n}$ (mm) | $t_{p,n}$ (mm) | $h_{w,n}$ (mm) | $t_{w,n}$ (mm) | $b_{f,n}$ (mm) | $t_{f,n}$ (mm) |
| 135–142, 145–149 | 980 | 24 | 625 | 15 | 150 | 25 |
| 150–151, 191–192 | 630 | 27.5 | 328 | 15 | 120 | 22 |
| 152–153 | 295 | 24 | - | - | - | - |
| 154–155 | 780 | 21 | 280 | 14 | 150 | 20 |
| 156–157 | 930 | 21 | 740 | 15.5 | 185 | 18 |
| 158–183 | 930 | 21 | 445 | 14 | 140 | 22 |
| 195 | 3370 | 30 | - | - | - | - |
| 197, 202–203, 208–209, 214 217, 219 | 465 | 23 | - | - | - | - |
| 198, 201, 204, 207, 210, 213 | 785 | 23 | 320 | 20 | - | - |
| 199–200, 205–206, 211–212 | 900 | 23 | 320 | 20 | - | - |
| 216, 218, 220 | 735 | 23 | 200 | 15 | - | - |

*Appendix A.2. Distribution of SWBM and VWBM*

The maximum SWBM throughout the ship's design life can be obtained according to IACS specifications [37].

$$M_{s0} = \begin{cases} -0.065C_wL^2B(C_B+0.7)(\text{sagging}) \\ C_wL^2B(0.1225-0.015C_B)(\text{hogging}) \end{cases} (\text{kNm}) \tag{A2}$$

where $B$ is the molded breadth and $C_B$ represents the block coefficient; $C_w$ is the wave coefficient, which can be determined by

$$C_w = \begin{cases} 10.75-((300-L)/100)^{3/2}, & 100 < L \le 300 \\ 10.75, & 300 < L \le 350 \\ 10.75-((L-350)/150)^{3/2}, & L > 350 \end{cases} \tag{A3}$$

The distribution of the extreme values of SWBM over a known period $T$ can be approximated by the Gumbel distribution.

$$F(x) = \exp\{-\exp[-\alpha_s(x-\mu_s^*)]\} \tag{A4}$$

where parameters $\mu_s^*$ and $\alpha_s$ are given by

$$\mu_s^* = \begin{cases} M_{s0}\cdot\sqrt{\frac{\ln(\nu_sT)}{\ln(\nu_sT_0)}}(\text{sagging}) \\ M_{s0}\cdot\frac{\ln(\nu_sT)}{\ln(\nu_sT_0)}(\text{hogging}) \end{cases} \tag{A5}$$

and

$$\alpha_s = \begin{cases} \frac{2\sqrt{\ln(\nu_sT)\cdot\ln(\nu_sT_0)}}{M_{s0}}(\text{sagging}) \\ \frac{\ln(\nu_sT_0)}{M_{s0}}(\text{hogging}) \end{cases} \tag{A6}$$

where $T_0$ represents the ship's design life; $\nu_s$ is the average incidence of each working condition.

The maximum VWBM can be defined by the IACS specification [37].

$$M_{w0} = \begin{cases} -0.11C_wL^2B(C_B+0.7) & (\text{sagging}) \\ 0.19C_wL^2BC_B & (\text{hogging}) \end{cases} (\text{kNm}) \tag{A7}$$

Similarly to the SWBM, the Gumbel distribution can also describe the extreme values of VWBM, expressed as follows

$$F(x) = \exp\{-\exp[-\alpha_w(x - \mu_w^*)]\} \quad (A8)$$

where parameters $\mu_s^*$ and $\alpha_s$ are defined by

$$\mu_w^* = M_{w0} \cdot \frac{\ln(\nu_w T)}{\ln(\nu_w T_0)} \quad (A9)$$

and

$$\alpha_w = \frac{\ln(\nu_w T_0)}{M_{w0}} \quad (A10)$$

## References


1. Sliva, L.A.R.; Torii, A.J.; Beck, A.T. Hyperstatic and redundancy thresholds in truss topology optimization considering progressive collapse due to aleatory and epistemic uncertainties. *Prob. Eng. Mech.* **2023**, *71*, 103384.
2. Zaitseva, E.; Levashenko, V.; Rabcan, J. An efficient curved inverse-shell element for shape sensing and structural health monitoring of cylindrical marine structures. *Ocean Eng.* **2023**, *229*, 108868.
3. Kang, R.; Zhang, Q.; Zeng, Z.; Enrico, Z.; Li, X.Y. Measuring reliability under epistemic uncertainty: Review on non-probabilistic reliability metrics. *Chin. J. Aeronaut.* **2016**, *29*, 571–579. [CrossRef]
4. Lee, C.H.; Kim, Y. Probabilistic flaw assessment of a surface crack in a mooring chain using the first- and second-order reliability method. *Mar. Struct.* **2019**, *63*, 1–15. [CrossRef]
5. Homaei, F.; Najafzadeh, M. A reliability-based probabilistic evaluation of the wave-induced scour depth around marine structure piles. *Ocean Eng.* **2020**, *196*, 106818. [CrossRef]
6. Okoro, A.; Khan, F.; Ahmed, S. An Active Learning Polynomial Chaos Kriging metamodel for reliability assessment of marine structures. *Ocean Eng.* **2021**, *235*, 109399. [CrossRef]
7. Xu, C.M.; Teixeira, A.P.; Soares, G.C. IACS-CSR buckling strength requirement-based response surface approach for reliability assessment of stiffened panels. *Ocean Eng.* **2024**, *302*, 117299. [CrossRef]
8. Meng, D.; Yang, S.; Yang, H.; De Jesus, A.M.P.; Correia, J.; Zhu, S.P. Intelligent-inspired framework for fatigue reliability evaluation of offshore wind turbine support structures under hybrid uncertainty. *Ocean Eng.* **2024**, *307*, 118213. [CrossRef]
9. Ben-Haim, Y.; Elishakoff, I. Non-Probabilistic Models of Uncertainty in the Nonlinear Buckling of Shells With General Imperfections: Theoretical Estimates of the Knockdown Factor. *J. Appl. Mech.* **1989**, *56*, 403–410. [CrossRef]
10. Wang, B.Y.; Zhang, X.Y.; Zhao, Y.G. Third moment method for reliability analysis with uncertain moments characterized as interval variables. *Struct. Saf.* **2024**, *111*, 102499. [CrossRef]
11. Tretiak, K.; Schollmeyer, G.; Ferson, S. Neural network model for imprecise regression with interval dependent variables. *Neural Netw.* **2023**, *161*, 550–564. [CrossRef] [PubMed]
12. Bai, Y.C.; Han, X.; Jiang, C.; Bi, R.G. A response-surface-based structural reliability analysis method by using non-probability convex model. *Appl. Math.* **2014**, *38*, 3834–3847. [CrossRef]
13. Maurya, S.; Tiwari, G. A convex modelling based reliability analysis of rock structures with limited data of inputs modelled via alternative uncertainty models: Application for rock slopes. *Comput. Geotech.* **2024**, *175*, 106674. [CrossRef]
14. Tango, Z.; Li, W.Q.; Li, Y. Discretization Analysis Method of Hybrid Reliability Based on Evidence Theory. *Math. Probl. Eng.* **2018**, *2018*, 9046708. [CrossRef]
15. Zhang, Z.; Jiang, C.; Wang, G.G.; Han, X. First and second order approximate reliability analysis methods using evidence theory. *Reliab. Eng. Syst. Saf.* **2015**, *137*, 40–49. [CrossRef]
16. Schobi, R.; Sudret, B. Structural reliability analysis for p-boxes using multi-level meta-models. *Probab. Eng. Mech.* **2017**, *48*, 27–38. [CrossRef]
17. Schobi, R.; Sudret, B. Uncertainty propagation of p-boxes using sparse polynomial chaos expansions. *J. Comput. Phys.* **2017**, *339*, 307–327. [CrossRef]
18. Faes, M.G.R.; Daub, M.; Marelli, S.; Patelli, E.; Beer, M. Engineering analysis with probability boxes: A review on computational methods. *Struct. Safe.* **2021**, *93*, 102092. [CrossRef]
19. Zhang, K.; Chen, N.; Liu, J.; Yin, S.; Beer, M. An efficient meta-model-based method for uncertainty propagation problems involving non-parameterized probability-boxes. *Reliab. Eng. Syst. Saf.* **2023**, *238*, 109477. [CrossRef]
20. Li, J.; Jiang, C. A novel imprecise stochastic process model for time-variant or dynamic uncertainty quantification. *Chin. J. Aeronaut.* **2023**, *35*, 255–267. [CrossRef]
21. Xiao, Z.; Zhang, Q.; Zhang, Z.; Bai, W.; Liu, H. A collaborative quasi-Monte Carlo uncertainty propagation analysis method for multiple types of epistemic uncertainty quantified by probability boxes. *Struct. Multidiscip. Optim.* **2023**, *66*, 109. [CrossRef]
22. Jiang, C.; Zheng, J.; Han, X. Probability-interval hybrid uncertainty analysis for structures with both aleatory and epistemic uncertainties: A review. *Struct. Multidiscip. Optim.* **2018**, *57*, 2485–2502. [CrossRef]

23. Wang, P.; Zhou, H.; Hu, H.; Zhang, Z.; Li, H. A novel method for reliability analysis with interval parameters based on active learning Kriging and adaptive radial-based importance sampling. *Int. J. Numer. Methods Eng.* **2022**, *123*, 3264–3284. [CrossRef]
24. Hu, H.; Deng, M.; Sun, W.; Li, J. A structural reliability analysis method under non-parameterized P-box based on double-loop deep learning models. *Struct. Multidiscip. Optim.* **2024**, *67*, 148. [CrossRef]
25. Haftka, R.T.; Nachalas, J.A.; Watson, L.T.; Rizzo, T.; Desai, R. Two-point constraint approximation in structural optimization. *Comput. Methods Appl. Mech. Eng.* **1987**, *60*, 289–301. [CrossRef]
26. Starnes, J.H.; Haftka, R.T. Preliminary Design of Composite Wings for Buckling, Strength, and Displacement Constraints. *J. Aircr.* **1979**, *16*, 564–570. [CrossRef]
27. Wang, L.P.; Grandhi, R.V. Improved Two-Point Function Approximations for Design Optimization. *AIAA J.* **1995**, *33*, 1720–1727. [CrossRef]
28. Grandhi, R.V.; Wang, L.P. Reliability-based structural optimization using improved two-point adaptive nonlinear approximations. *Finite Elem. Anal. Des.* **1998**, *29*, 35–48. [CrossRef]
29. Keshtegar, B.; Chakraborty, S. A hybrid self-adaptive conjugate first order reliability method for robust structural reliability analysis. *Appl. Math. Model.* **2018**, *53*, 319–332. [CrossRef]
30. Huang, Z.L.; Jiang, C.; Zhou, Y.S.; Zheng, J.; Long, X.Y. Reliability-based design optimization for problems with interval distribution parameters. *Struct. Multidisc. Optim.* **2017**, *55*, 513–528. [CrossRef]
31. Zhang, S.M.; Khan, I. Buckling and ultimate capability of plates and stiffened panels in axial compression. *Mar. Struct.* **2009**, *22*, 791–808. [CrossRef]
32. Kim, D.K.; Lim, H.L.; Kim, M.S.; Hwang, O.J.; Park, K.S. An empirical formulation for predicting the ultimate strength of stiffened panels subjected to longitudinal compression. *Ocean Eng.* **2017**, *140*, 270–280. [CrossRef]
33. Sun, H.H.; Bai, Y. Time-variant reliability assessment of FPSO hull girders. *Mar. Srtuct.* **2003**, *16*, 219–253. [CrossRef]
34. IACS. *Technical Report for CSR-H External Release on 2018-03-26*; International Association of Classification Societies: London, UK, 2018.
35. Paik, J.K.; Mansour, A.E. simple formulation for predicting the ultimate strength of ships. *J. Mar. Sci. Technol.* **1995**, *1*, 52–56. [CrossRef]
36. Paik, J.K.; Thayamballi, A.K. An empirical formulation for predicting the ultimate compressive strength of stiffened panels. In Proceedings of the 7th International Offshore and Polar Engineering Conference, Honolulu, HI, USA, 25–30 May 1997.
37. IACS. *Common Structural Rules for Double Hull Oil Tankers*; International Association of Classification Societies: London, UK, 2012.

Journal of
*Marine Science and Engineering*

MDPI

*Article*

# Study on Strain Field Reconstruction Method of Long-Span Hull Box Girder Based on iFEM

**Guocai Chen [1,2], Xueliang Wang [1,2,*], Nan Zhao [1,2], Zhentao Jiang [1,2], Fei Li [1,2], Haozheng Chen [1,2], Pengyu Wei [1,2] and Tao Zhang [1,2]**

1 China Ship Scientific Research Center, Wuxi 214082, China; chenguocai2686@163.com (G.C.)
2 National Key Laboratory of Ship Structural Safety, Wuxi 214082, China
* Correspondence: wangxl@cssrc.com.cn

**Abstract:** The box girder's condition significantly impacts the safety and overall performance of the entire ship because it is the primary stress component of the hull construction. This work used experimental research on the long-span hull box girder based on IFEM (Inverse Finite Element Method) technology to ensure the structural safety of the hull box girder. Due to the limitations of conventional experiments in this technical field, such as their reliance on finite element data and lack of input from physical tests, numerous research methods combining the strain sensing data from physical tests with the strain data from virtual sensors were conducted. The strain fields of the top plate, side plate, and bottom plate were each reconstructed in turn, and the verifier measuring points in the physical model test were used to assess the accuracy of the reconstruction results. The findings demonstrate that the top plate, side plate, and bottom plate reconstructions had relative errors of 0.24–7.86%, 0.75–8.13%, and 3.31–2.52%, respectively. This enables the reconstruction of the strain field of the long-span hull box girder using physical test data and promotes the use of iFEM technology in the field of structural health monitoring of large marine structures.

**Keywords:** iFEM; box girder; combination method of reality and excess; strain field reconstruction; structural monitoring

## 1. Introduction

Due to their small weight and excellent strength, box girders are frequently employed in ships, buildings, bridges, and other technical applications. The box girder, which is used in ship construction, essentially defines the entire longitudinal bearing capacity of the ship and increases its longitudinal strength. Long-term cyclic loads and short-term high loads can cause various types of deformations in ships, including bending, torsional deformations, and fatigue damage. Long-term cyclic loads, such as those induced by repeated wave stresses, can lead to material fatigue, gradually weakening the structural integrity. On the other hand, short-term high loads, such as those resulting from impacts or extreme weather events, may cause localized buckling, plastic deformation, or even sudden structural failure. These deformations not only compromise the structural integrity of the ship, but also pose significant risks to its durability and safety. Therefore, it is crucial to account for these factors in ship design and maintenance. Setting up a structural health-monitoring (SHM) system for the hull box girder to gather local or global strain data and perform precise structural health management is a crucial step in efficiently ensuring the safety of the structure [1,2]. Measurement of the high-precision strain field of fused and rebuilt structures using discrete structural strain information has long been a hot topic for researchers both domestically and internationally because it serves as a key indicator of structural response.

Four popular strain field reconstruction techniques are being used both domestically and internationally: the modal method [3–5], KO displacement theory [6,7], artificial intelligence [8], and the iFEM method [9,10]. The fundamental tenet of the modal technique is that structural deformation is viewed as a linear combination of several modes and that the precision of modal analysis has a significant bearing on the accuracy of reconstruction. Because Ko displacement theory is derived from the deflection curve, it can only be used to reconstruct unidirectional structural deformation [11]. The artificial neural network model serves as the foundation for most artificial intelligence techniques. The deformation reconstruction model has strong universality and a weak correlation with the material properties, structural parameters, and distribution form of the environmental load of the measured object, but the precision of the reconstruction displacement is greatly influenced by the richness of the load set used to train the network [12]. IFEM builds the least square error function between the measured strain value and the theoretical value and combines the shape and size information. The potential for using iFEM in engineering is very strong because it can do away with the restrictions of conventional reconstruction techniques and ignore the material qualities and load of the structure. Scholars domestically and internationally have conducted extensive research on plates, girders, shells, and other structures since Adnan Tessler and Spangler [13] first officially presented the concept and theory of iFEM in 2003. This research has further confirmed the robustness and viability of iFEM in the fields of structural health monitoring of wings and hulls, as well as the reconstruction of composite deformation fields [14,15].

Currently, the majority of iFEM technology research in the fields of ship and ocean engineering is conducted abroad. Adnan Kefal [16] et al. made the initial demonstration of iFEM's suitability for use in managing the health and safety of offshore structures in 2015. It is believed that the iFEM method is a promising technology for precise real-time monitoring of offshore structures after theoretical verification of the stiffened plates, such as the quadrilateral plate bearing bending force and the side parts of typical longitudinal and transverse frame tankers bearing bending load. Adnan Kefal [17] et al. used the iFEM approach to monitor the displacement and tension of the midship of a conventional chemical tanker for the first time in 2016. However, a long barge with the same cross-section as a conventional chemical tanker was simplified by employing the iQS4 element as opposed to the original model of a chemical tanker. In 2016, Adnan Kefal [9] et al. used the iQS4 element to simulate the typical parallel intermediate of the Panamanian container ship. Using the simulated sensor strains obtained under three different conditions—pure vertical bending, pure horizontal bending, and pure torsion—they then made three different iFEM analyses of the parallel intermediate. Adnan Kefal [18] used the FBG (fiber Bragg grating) strain sensing technology in 2018 to reconstruct the real-time three-dimensional displacement and stress of the overturned bulk carrier. Adnan Kefal's research completely confirmed the viability and robustness of iFEM, but it was based on numerical simulation, meaning that no actual physical model testing was conducted and the strain output of the finite element model was used to represent the "experimental" strain measurement (that is, in-situ strain data or virtual sensor strain). Putranto T took the hull box girder as the research object, and ultimate strength predictions obtained from equivalent single-layer (ESL) approach were compared to the full three-dimensional finite element method (3D FEM) and the international association of classification societies (IACS) incremental-iterative method. The comparison between different methods was provided in terms of longitudinal bending moment and cross-sectional stress distribution. Overall, the ESL approach yields good agreement compared to the 3D FEM results in predicting the ultimate strength of a ship hull girder while providing up to 3 times the computational efficiency and ease of modeling [19,20].

Domestic research in the fields of naval architecture and marine engineering was slow to begin. In 2023, Yan Hongsheng [21] and others applied iFEM to the deformation reconstruction of a ship-stiffened plate structure, but this falls under the two-dimensional structure reconstruction category. Hu Mingyue [14] and colleagues performed the reconstruction of a three-dimensional wallboard structure's deformation field in 2022. The issue of the structure's discontinuous internal rotation angle under the over-constraint condition was resolved by combining sub-region division and the inverse finite element method, and the statically indeterminate structure's full-field deformation reconstruction was realized using measured strain data. However, in the process of experimental verification, there were considerably fewer physical measurement points than under the simulation settings, and it was not possible to reconstruct the strain field using the physical test data as input. In 2023, Qingfeng Zhu proposed a method for strain reconstruction in stiffened ship panels, employing optical fiber sensors and the strain function–inverse finite element principle (SF-iFEM), addressing the crucial need for monitoring the health of ship structures [22]. In 2024, Pengyu Wei employed the inverse finite element method to derive the deformation field of the ship structure in real time using sensor strain data. The deformation field data obtained based on the iFEM algorithm were converted into general visualization data conducive to interpretation within virtual reality (VR) applications. Lastly, a digital twin software tool was built to enable synchronous responses and interactions between the virtual scene and the physical scene [23]. The potential benefits of iFEM in the realm of engineering application can be tested by reconstructing the strain field using the measured data from the physical model test as input. Additionally, the strain field reconstruction based on the results of the physical model test is more trustworthy and significant for engineering applications.

Based on this, this paper conducts the first study on the long-span hull box girder based on iFEM technology. The research direction of combining a significant amount of physical test strain sensing data with virtual sensor strain data is explored due to the phenomenon that the traditional test in this technical field heavily relies on finite element data, there are few experimental measurement data., and the reconstruction results have a low degree of matching with the actual strain field of the physical model. The strain field of long-span hull box girders using physical test data is realized, laying the groundwork for structural health monitoring of massive marine structures and the application. The strain field of the long-span hull box girder is reconstructed according to the structural form of the hull box girder and according to the top surface, side surface, and bottom surface, and the accuracy of the reconstruction results is verified by physical verification measuring points.

## 2. Inverse Finite Element Formulation for Shells

### 2.1. Constructing Inverse Shell Element

In order to adapt to different structural forms and physical models, different inverse elements need to be constructed. At present, iCS8 and iQS4 are often used in the reconstruction of deformation field of offshore structures. iCS8 is suitable for constructing inverse shell elements of curved structures such as submarines and cylinders. Because the strain studied in this paper is in a linear changing stage; the top, side, and bottom plates are flat; and the influence of shear deformation on the structure is considered, the strain field is rebuilt using Mindlin plate theory and iQS4 as the inverse shell element [17]. The local coordinate system $(x, y, z)$ of a plate element is constructed, as seen in Figure 1, with the coordinate origin at the centroid of the neutral surface of the plate element. $u$, $v$, and $w$ indicate the displacements in the directions of $x$, $y$, and $z$ in the local coordinate system. The rotation angles about the axes of $x$, $y$, and $z$ are $\theta_x$, $\theta_y$, and $\theta_z$, respectively. The four nodes of the element are points 1, 2, 3, and 4, and the thickness of the element is $2h$.

$$\begin{cases} u = u_0 + z\theta_{y_0} \\ v = v_0 - z\theta_{x_0} \\ w = w_0 \end{cases} \tag{1}$$

where the distributions of $u_0$, $v_0$, and $u_0$ represent the displacements of points in directions $x, y, z$; $\theta_{x_0}$ and $\theta_{y_0}$ denote the rotation angles of a point on the median plane around the $x$-axis and $y$-axis directions, where the displacement $w$ in the $z$-axis direction is unchanged along the thickness direction, $z \in [-h, +h]$.

$$N_i = \frac{1}{4}(1 + x_i x)(1 - y_i y), i = 1,2,3,4 \tag{2}$$

$$\boldsymbol{u}_i^{\mathrm{e}} = [u_i\ v_i\ w_i\ \theta_{xi}\ \theta_{yi}\ \theta_{zi}]^T\ i = 1,2,3,4 \tag{3}$$

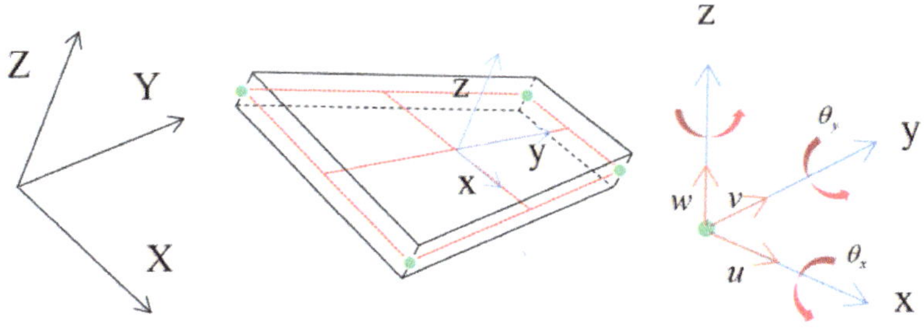


**Figure 1.** Four-node inverse shell element iQS4; (**a**) iQS4 element showing global and local coordinate systems; (**b**) nodal DOF in the local coordinate system xyz.

The displacement matrix of the four-node element can be expressed as:

$$\boldsymbol{u}^{\mathrm{e}} = [u_1\ u_2\ u_3\ u_4]^T \tag{4}$$

As stated in Formulas (5)–(7), the theoretical strain of a box girder structure can be broadly divided into linear combinations of in-plane tension and compression strain, bending strain, and shear strain.

$$e(u^e) = \begin{Bmatrix} e_1 \\ e_2 \\ e_3 \end{Bmatrix} = \begin{Bmatrix} \frac{\partial u}{\partial x} \\ \frac{\partial v}{\partial y} \\ \frac{\partial u}{\partial y} + \frac{\partial v}{\partial x} \end{Bmatrix} \tag{5}$$

$$k(u^e) = \begin{Bmatrix} k_1 \\ k_2 \\ k_3 \end{Bmatrix} = \begin{Bmatrix} \frac{\partial \theta_y}{\partial x} \\ \frac{\partial \theta_x}{\partial y} \\ \frac{\partial \theta_x}{\partial x} + \frac{\partial \theta_y}{\partial y} \end{Bmatrix} \tag{6}$$

$$s(u^e) = \begin{Bmatrix} s_1 \\ s_2 \end{Bmatrix} = \begin{Bmatrix} \frac{\partial w}{\partial x} + \theta_y \\ \frac{\partial w}{\partial y} - \theta_x \end{Bmatrix} \tag{7}$$

Additionally, the relationship between theoretical strain and the element node displacement matrix can be established through the partial derivative matrix of shape function. As shown in Formula (8), the surface strain $\varepsilon_b$ of the box girder structure can be expressed as partial derivative matrices $\boldsymbol{B}^{\mathrm{m}}$ and $\boldsymbol{B}^{\mathrm{k}}$ of the four-node inverse shell element shape function and element node displacement matrix $u^e$.

$$\varepsilon_b = e(\boldsymbol{u}^{\mathrm{e}}) + zk(\boldsymbol{u}^{\mathrm{e}}) = \boldsymbol{B}^{\mathrm{m}}\boldsymbol{u}^{\mathrm{e}} + z\boldsymbol{B}^{\mathrm{k}}\boldsymbol{u}^{\mathrm{e}} \tag{8}$$

The partial derivative matrix $B^s$ of the four-node inverse shell element shape function and the element node displacement matrix $\boldsymbol{u}^{\mathrm{e}}$ can both be used to define the transverse shear strain $\varepsilon_s$, as shown in Formula (9).

$$\varepsilon_s = s(\boldsymbol{u}^{\mathrm{e}}) = \boldsymbol{B}^{\mathrm{s}}\boldsymbol{u}^{\mathrm{e}} \tag{9}$$

The literature [17] illustrates the precise calculation method and theory of the shape function partial derivative matrices $\boldsymbol{B}^{\mathrm{m}}$, $\boldsymbol{B}^{\mathrm{k}}$, and $\boldsymbol{B}^{\mathrm{s}}$.

*2.2. Data Input*

Using physical strain sensors and virtual strain sensors to measure the strain of the upper and lower surfaces of the box girder, respectively, the strain of the upper surface is $\left(\varepsilon_{xx}^{+}, \varepsilon_{yy}^{+}, s_{xy}^{+}\right)_j$ and the strain of the lower surface is $\left(\varepsilon_{xx}^{-}, \varepsilon_{yy}^{-}, s_{xy}^{-}\right)_j$, As shown in Figure 2. $j = 1,2,3 \cdots n$, $n$ is the number of strain sensors in the element, "+"and "−"respectively represent the upper and lower surfaces of the structure, and $2h$ is the plate thickness of the box girder, which can be expressed by the measured strain of the upper and lower surfaces in Equations (10) and (11). Virtual strain sensor refers to the technical means of using the numerical analysis method to give the strain response of structural target position under different loading conditions instead of the physical strain sensor.

$$e_j^{\mathrm{c}} = \frac{1}{2}(\varepsilon_j^{+} + \varepsilon_j^{-})\, j = 1,2,3 \cdots n \tag{10}$$

$$k_j^{\mathrm{c}} = \frac{1}{2h}(\varepsilon_j^{+} - \varepsilon_j^{-})\, j = 1,2,3 \cdots n \tag{11}$$

$$\varepsilon_j^{+} = \left[\varepsilon_{xx}^{+}\ \varepsilon_{yy}^{+}\ s_{xy}^{+}\right]_j^T;\ \varepsilon_j^{-} = \left[\varepsilon_{xx}^{-}\ \varepsilon_{yy}^{-}\ s_{xy}^{-}\right]_j^T$$

where: superscript $c$ stands for the discrete strain detected in the test, superscript $\boldsymbol{e}_j^{\mathrm{c}}$ stands for the tensile and compressive strain of the plate neutral surface throughout the test, and superscript $\boldsymbol{k}_j^{\mathrm{c}}$ stands for the bending strain of the plate neutral surface. Since the shear strain $\boldsymbol{s}_j^{\mathrm{c}}$ is significantly smaller than the tensile-compressive strain $\boldsymbol{e}_j^{\mathrm{c}}$, it is omitted during measurements for simplicity and subsequently determined using a proportional relationship.

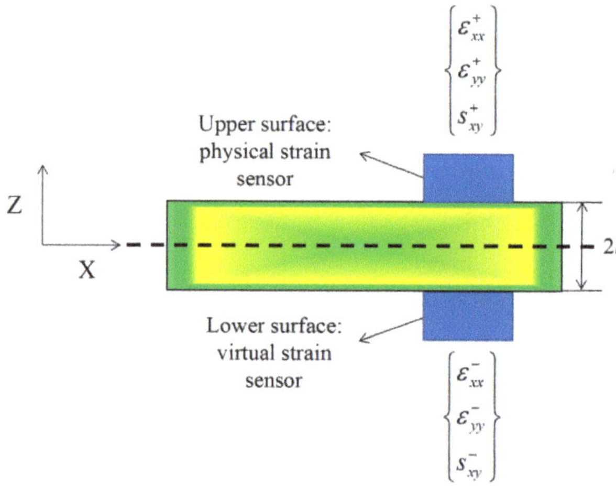


**Figure 2.** Discrete surface strain measured by strain flower in iQS4 element.

*2.3. Least Squares Error Function*

Between the measured strain and the theoretical strain, the least square error function is built. The theoretical displacement value of each element node is obtained when the partial derivative of the error function with respect to the displacement of the element node is equal to 0, and it is assumed that the displacement distribution of each element node at this time corresponds to the actual displacement distribution in the measured strain state at this time.

$$\phi_e(\boldsymbol{u}^e) = w_e\|e(\boldsymbol{u}^e) - \boldsymbol{e}^c\|^2 + w_k\|k(\boldsymbol{u}^e) - \boldsymbol{k}^c\|^2 + w_s\|s(\boldsymbol{u}^e) - \boldsymbol{s}^c\|^2 \tag{12}$$

Among them, $e(\boldsymbol{u}^e)$, $k(\boldsymbol{u}^e)$, $s(\boldsymbol{u}^e)$ are the theoretical tension, compression, bending, and shear strains of the element represented by neutral surface displacement field; $\boldsymbol{e}^c$, $\boldsymbol{k}^c$, and $\boldsymbol{s}^c$ are the tension, compression, bending, and shear strains of the neutral surface derived from the surface strain information obtained by simulation test or strain gauge measurement; and $w_e$, $w_k$, and $w_s$ are dimensionless weighting coefficients, which are related to the strain of each section and control the strength relationship between the theoretical neutral surface strain and the actual neutral surface strain. If the strain $\boldsymbol{e}^c$, $\boldsymbol{k}^c$, and $\boldsymbol{s}^c$ of the element can be obtained, the weighting coefficient can be $w_e = w_k = w_s = 1$. For the case of an actual lack of strain, such as the shear strain mentioned above, the accuracy and rationality of the result can be controlled by adjusting the corresponding coefficient [11,24]. After a comprehensive analysis of the structure and material of the box girder, $w_s = \lambda = 10^{-4}$ ($0 < \lambda \leq 1$, $\lambda$ is the penalty parameter) [25–27].

$$\|e(\boldsymbol{u}^e) - \boldsymbol{e}^c\|^2 = \frac{1}{n}\iint\limits_{A^e}\sum_{i=1}^{n}(e(\boldsymbol{u}^e)_i - \boldsymbol{e}_i^c)^2 dxdy \tag{13}$$

$$\|k(\boldsymbol{u}^e) - \boldsymbol{k}^c\|^2 = \frac{(2h)^2}{n}\iint\limits_{A^e}\sum_{i=1}^{n}(k(\boldsymbol{u}^e)_i - \boldsymbol{k}_i^c)^2 dxdy \tag{14}$$

$$\|s(\boldsymbol{u}^e) - \boldsymbol{s}^c\|^2 = \frac{1}{n}\iint\limits_{A^e}\sum_{i=1}^{n}(s(\boldsymbol{u}^e)_i - \boldsymbol{s}_i^c)^2 dxdy \tag{15}$$

where: $A^e$ is the area in the plate element; $n$ is the number of sensors in the element.

$$\frac{\partial\phi_e(\boldsymbol{u}^e)}{\partial\boldsymbol{u}^e} = \boldsymbol{k}^e\boldsymbol{u}^e - \boldsymbol{f}^e = 0 \tag{16}$$

The Formula (16) can be obtained:

$$\boldsymbol{k}^e\boldsymbol{u}^e = \boldsymbol{f}^e \tag{17}$$

$$\boldsymbol{k}^e = \iint\limits_{A^e}[(\boldsymbol{B}^m)^T\boldsymbol{B}^m + (2h)^2(\boldsymbol{B}^k)^T(\boldsymbol{B}^k) + \lambda(\boldsymbol{B}^s)^T(\boldsymbol{B}^s)]dxdy \tag{18}$$

$\boldsymbol{f}^e$ matrix consists of $\boldsymbol{B}^m, \boldsymbol{B}^k, \boldsymbol{B}^s$ matrices and their corresponding weighting coefficients $w_e$, $w_k$, and $w_s$, which are also related to the number of strain sensors $n$ and the thickness of the cell plate $h$. $w_e = w_k = 1$, $w_s = \lambda = 10^{-4}$. The specific form is as follows:

$$\boldsymbol{f}^e = \frac{1}{n}\iint\limits_{A^e}[(\boldsymbol{B}^m)^T\boldsymbol{e}_i^c + (2h)^2(\boldsymbol{B}^k)^T\boldsymbol{k}_i^c + \lambda(\boldsymbol{B}^s)^T s_i^c]dxdy \tag{19}$$

Assemble the element matrix to create the entire matrix. The stiffness matrix equation of the element in the local coordinate system can be transformed into the stiffness matrix equation in the global coordinate system using the coordinate transformation matrix, and then the stiffness matrix equation of the discrete structure can be integrated to obtain the overall stiffness matrix equation of the overall structure. This procedure is analogous to the thought of finite elements [15].

$$\boldsymbol{KU} = \boldsymbol{F} \tag{20}$$

The specific solution and form are as follows:

$$K = \sum_{e=1}^{n_{el}} (T^e)^T k^e T^e \tag{21}$$

$$F = \sum_{e=1}^{n_{el}} (T^e)^T f^e \tag{22}$$

$$U = \sum_{e=1}^{n_{el}} (T^e)^T u^e \tag{23}$$

where $\boldsymbol{K}$ is the overall stiffness matrix, which is related to the position and number of discrete elements in the structure; $n_{el}$ is the total number of inverse finite elements; $\boldsymbol{U}$ is the overall structural displacement vector; $\boldsymbol{F}$ is the overall load matrix, which is related to the mid-plane strain of discrete elements in the structure; and $\boldsymbol{T}^e$ is the coordinate transformation matrix. The $\boldsymbol{k}$ matrix is a symmetric matrix that has nothing to do with the strain measurement value, but is related to the location of the cell node and the location of the strain measurement point in the inverse cell. The coefficient matrix $\boldsymbol{k}$ will be simplified to a positive definite matrix when combined with the cell boundary conditions, and the global displacement $\boldsymbol{U}$ of the cell node can be determined after inversion.

When the shear deformation of the plate element is ignored, the shape function partial derivative matrix and the element's node displacement matrix can be used to express it.

$$\varepsilon_b = e(u^e) + zk(u^e) = B^m u^e + zB^k u^e = \left(B^m + zB^k\right)u^e \tag{24}$$

If the structural shear strain of the plate element is taken into account, the penalty parameters can be suitably chosen based on the relationship between shear strain and tensile and compressive strain $\lambda$.

$$\begin{aligned} \varepsilon &= \mathrm{e}(\boldsymbol{u}^{\mathrm{e}}) + \mathrm{z}\mathrm{k}(\boldsymbol{u}^{\mathrm{e}}) + \lambda \mathrm{e}(\boldsymbol{u}^{\mathrm{e}}) = \boldsymbol{B}^{\mathrm{m}}\boldsymbol{u}^{\mathrm{e}} + \mathrm{z}\boldsymbol{B}^{\mathrm{k}}\boldsymbol{u}^{\mathrm{e}} + \lambda\boldsymbol{B}^{\mathrm{s}}\boldsymbol{u}^{\mathrm{e}} \\ &= \left(\boldsymbol{B}^{\mathrm{m}} + \mathrm{z}\boldsymbol{B}^{\mathrm{k}} + \lambda\boldsymbol{B}^{\mathrm{s}}\right)\boldsymbol{u}^{\mathrm{e}} \end{aligned} \tag{25}$$

$$\varepsilon(x,y) = \begin{bmatrix} \varepsilon_{xx} \\ \varepsilon_{yy} \\ \gamma_{xy} \end{bmatrix} = [\partial]\boldsymbol{u} = [\partial]\boldsymbol{N}\boldsymbol{q}^e = \boldsymbol{B}\boldsymbol{q}^e \tag{26}$$

For the assembly of discrete strain elements, the form function partial derivative matrix is multiplied by the element node displacement matrix to give the global strain $\varepsilon$.

## 3. Model Parameters

The model material of the hull box girder studied in this paper is Q235 steel, with a total length of 12.5 m, a width of 1.5 m, and a height of 0.8 m. The model is divided into two loading sections, two changeover portions, one test section, and two tooling occupation sections, which are symmetrically distributed along the length with the test section as the center. In numerical simulation, the thickness of the top plate is 5.55 mm, the other three plates are 5.75 mm, the elastic modulus is set to 205 GPa, and Poisson's ratio is 0.3. It should be noted that the test section is created thinner than the plates in the loading and transition sections during model processing to ensure that the test can be effectively controlled and damaged in the test section, and no longitudinal bone material is set. When the model boundary is set, two simple supports are set at the top plate to limit the vertical movement of the model, and different upward thrust forces are applied to the tooling loading section at the top of the model. During the test, there are two 4000 kN hydraulic cylinders at the bottom of the model, which work together on a horizontal cross bar, and the top of the model is fixed by a fixture, which contacts the top plate of the box girder in a semi-circular arc to realize the four-point bending loading test of the model. Figure 3 is a schematic front view of model loading. Table 1 displays the geometric characteristics and finite element material properties of the test model.

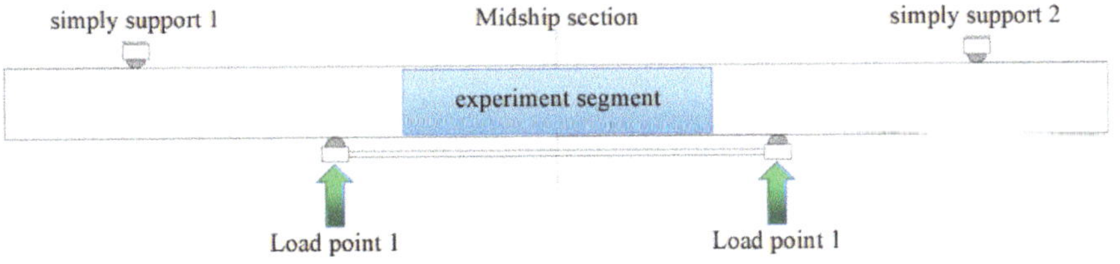


**Figure 3.** Schematic diagram of four-point bending of hull box girder.

**Table 1.** Geometric parameters and numerical simulation material properties of box girder.

| Box Girder Dimensions (mm) | | Division and Size of Box Girder Cabin (mm) | | Quantity (pcs) | Thickness of Plate and Shell in Test Section (mm) | | Elastic Modulus (GPa) | Poisson's Ratio |
|---|---|---|---|---|---|---|---|---|
| Length | 12,500 | Loading section | 1500 | 2 | Top plate | 5.55 | 205 | 0.3 |
| Width | 1500 | Changeover portion | 1500 | 2 | Port | 5.75 | 205 | 0.3 |
| Height | 800 | Test section | 4500 | 2 | Bottom plate | 5.75 | 205 | 0.3 |
| | | Tooling occupation section | 500 | 4 | Starboard | 5.75 | 205 | 0.3 |

Figures 4 and 5 present numerical simulations conducted using the finite element software Abaqus 2022. The primary goal of these simulations is to guide the placement of measurement points in the experimental model. The strain distribution on the port exhibits a trend from the top to the bottom, transitioning from a positive high-strain zone to a low-strain zone, and then to a negative high-strain zone. Accordingly, strain sensors are placed in each of these zones. Additionally, strain data from virtual sensors within the model are used to supplement the experimental data from the physical model. The hull box girder has two longitudinal ribs (three rows of areas) on the left and right sides, and five longitudinal ribs (six rows of areas) on the top and lower sides. By evaluating the model's longitudinal strain distribution law, it is discovered that the strain is symmetrical left and right along the middle longitudinal portion and symmetrical bow and stern along the middle transverse section; thus, the test scheme is established in portion 3.1 in accordance with this law.

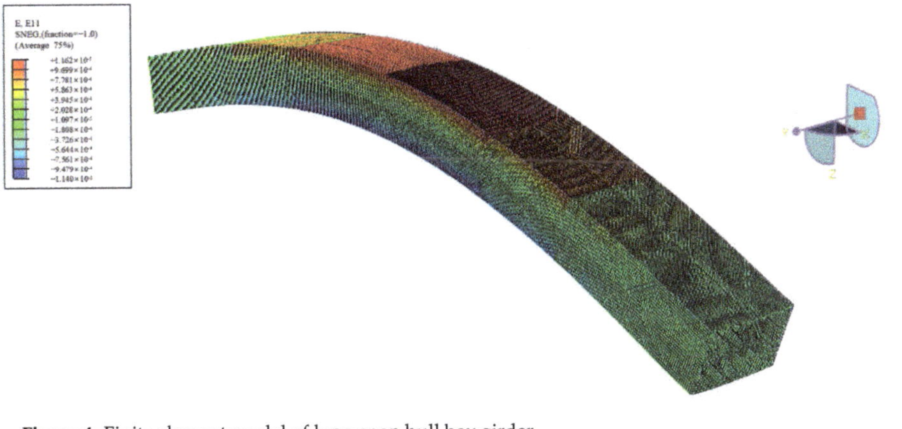

**Figure 4.** Finite element model of long-span hull box girder.

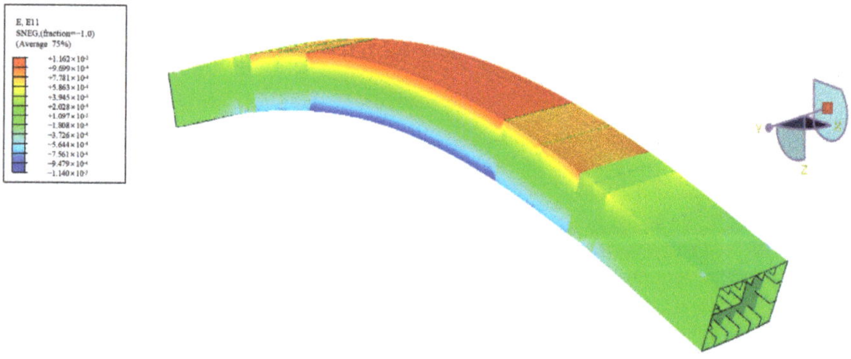

**Figure 5.** Nephogram of strain distribution in direction (longitudinal direction) of box girder.

## 4. Design of Test Scheme for Hull Box Girder

The experimental model adopts two kinds of sensor acquisition technologies: strain gauge and optical fiber. In this paper, only the acquisition results of strain gauge sensor are used as the input of strain field reconstruction, and the results of the optical fiber sensor are not collected. In this paper, the information acquisition process based on strain gauge is to connect the strain gauge and the acquisition instrument through a 9-core cable to realize the collection and transmission of structural strain information, and then reconstruct the strain field of the structure after processing the collected results. The main process of structural strain information acquisition is shown in Figure 6.

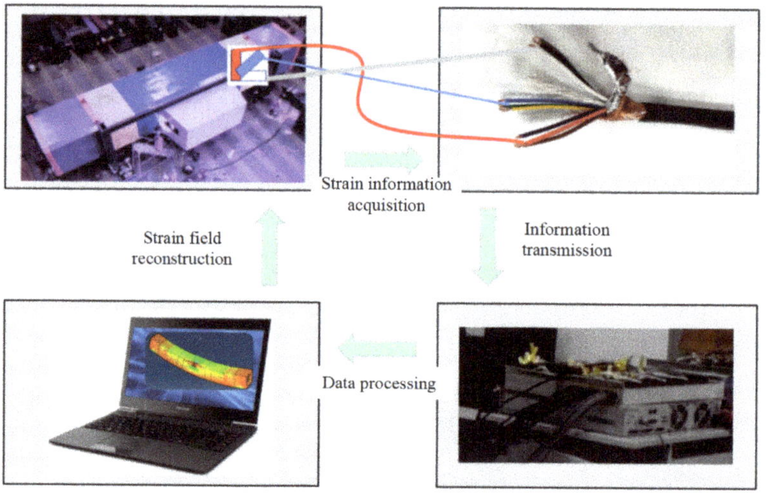


**Figure 6.** Test data acquisition–application process.

Under the condition of four-point bending, by analyzing the numerical simulation results, it is found that the distribution of strain field of box beam is symmetrical in the middle longitudinal section and symmetrical in the middle transverse section. Therefore, the distributed optical fiber sensor and strain sensor are set in two symmetrical areas along the longitudinal section, and the main purpose is to form mutual verification of the acquisition accuracy of the two sensing acquisition technologies. The collected data show that the accuracy of the two sensing acquisition technologies is high in the linear stage of the structure, but only the strain gauge acquisition technology is used for related research in this study. Additionally, as indicated in Figure 7, the box girder is separated into multiple compartments along the longitudinal direction in order to meet the test criteria.

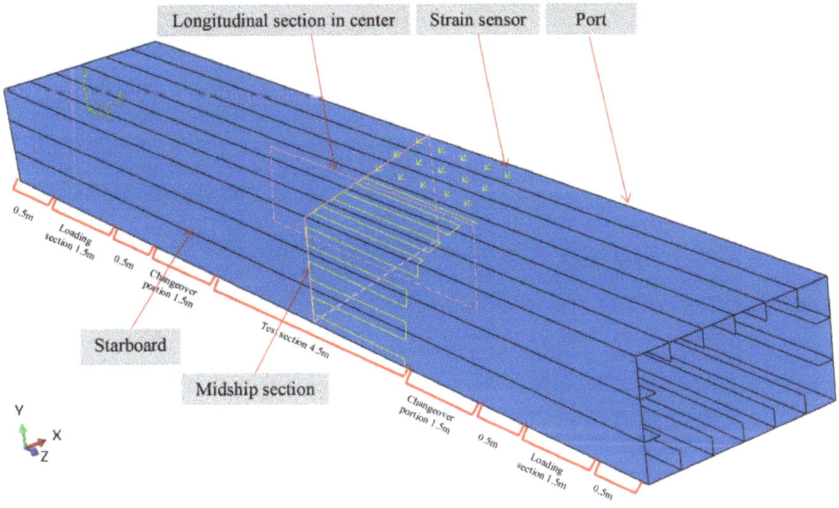


**Figure 7.** Schematic diagram of compartments division.

As shown in Figure 8, the layout area of the strain rosettes is symmetrical with the optical fiber along the longitudinal section, and 45 reconstructed physical measuring points are placed every 0.5 m along the centerline of the array board. Additionally, 12 additional strain rosettes and 6 unidirectional strain gauges are placed as verification points. Of the points, 45 are used for reconstructing the strain field, while the 15 validation points are solely for verifying the reconstruction accuracy and do not participate in the strain field reconstruction. Figure 9 is the sensor layout diagram of the test site.

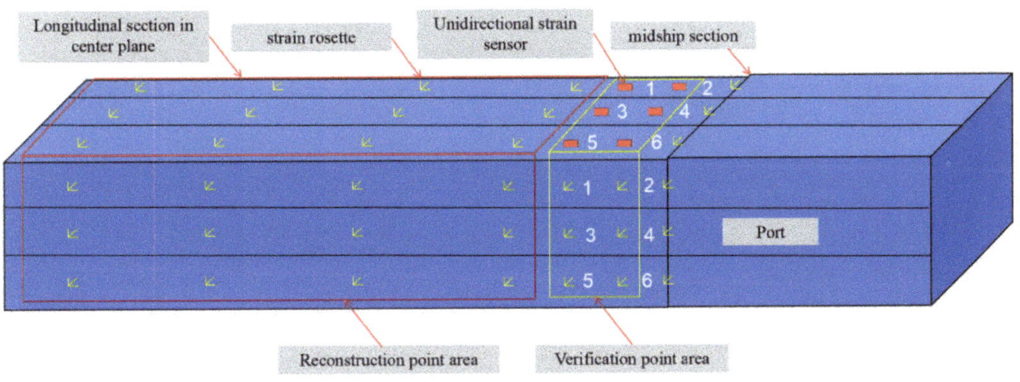


**Figure 8.** Schematic diagram of strain sensor arrangement.

The long-span hull's box girder underwent a four-point bending test to imitate the mid-arch when the ship was sailing against the waves. The semi-circular loading block was in line contact with the surface of the box girder, the loading section above the box girder was fixed by tooling, and the two oil cylinders at the bottom were loaded simultaneously. Figure 10 shows the schematic diagram of the test loading, and Table 2 shows the loading conditions. The test site is shown in Figure 11.

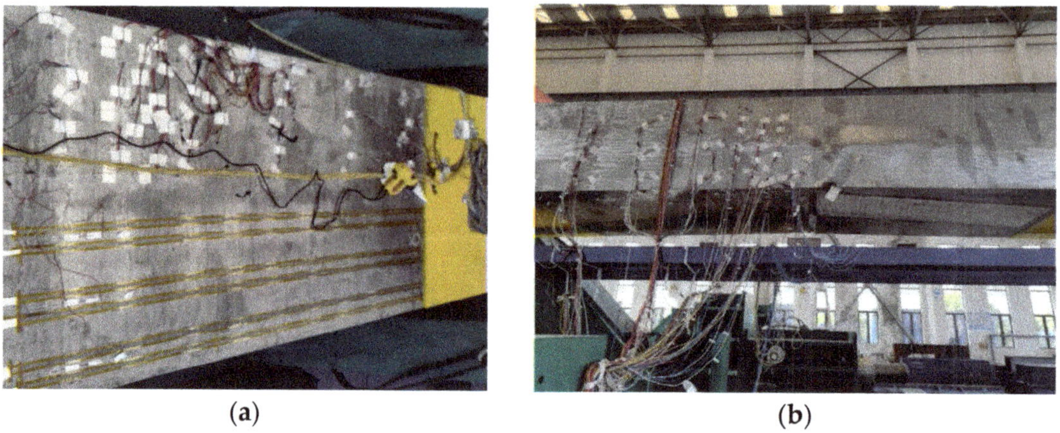

Figure 9. Layout of physical sensors on the test site; (a) layout of strain sensor on top surface; (b) layout diagram of shipboard physical sensor.

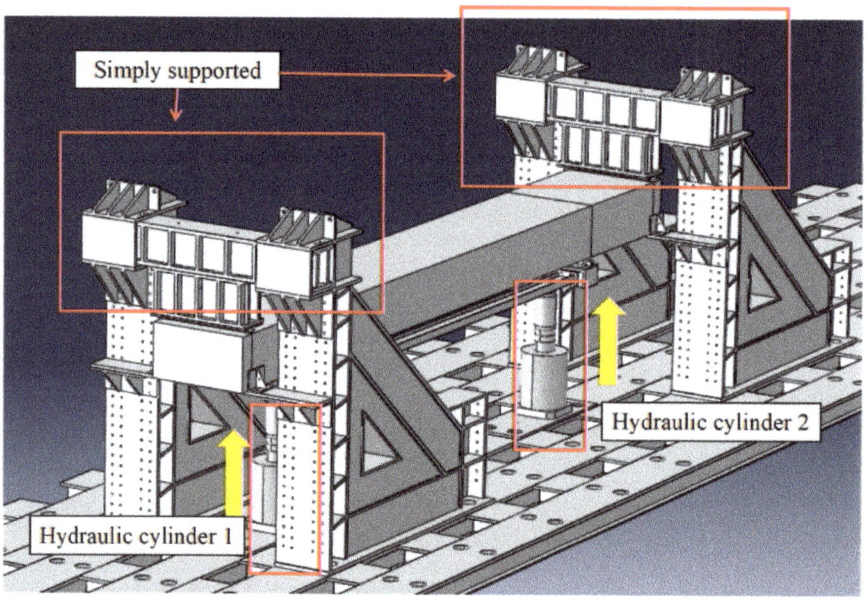


Figure 10. Loading diagram of four-point bending test of long-span box girder.

Table 2. Box girder loading condition table.

| Loading Type | Load Size (kN) | Number of Replication |
|---|---|---|
| Four-point bending | 0, 200, 400, 600, ...1400, 1600, 1700... | 1 |

**Figure 11.** Test loading site of long-span box girder.

## 5. Discussion

### 5.1. Analysis of Test Data

The experimental data under 1200 kN working conditions were used in this research as an example to show the reconstruction, and the linear stage reconstruction method was the same under various working conditions. The top plate was in tension, the bottom plate was in compression, and the neutralization axis was close to the centerline of the side plate when the long-span box girder was bent at four points. The whole trend fits the "pull positive pressure negative" strain pattern. The trend is also consistent with the strain distribution characteristics of long-span box girders, as shown in Figure 8, as the number of measuring points was arranged from top surface to side surface to bottom surface from left to right, with five measuring points in one row, a total of nine rows, and nine local fluctuations.

The relative error of the measured data at the verification sites is assessed in accordance with the reconstruction results in order to confirm the accuracy of the strain field reconstruction for long-span box girders. One can easily and intuitively represent the relative error between the strain reconstruction value and the measured value of the verification point using $PD(\varepsilon)$. Among them, $\varepsilon_{L,i}^{iFEM}(i = 1, \quad 2, \quad 3\ldots18)$ represents the strain reconstruction value of each verification point along the longitudinal direction of the box girder, and $\varepsilon_{L,i}^{VP}(i = 1, \quad 2, \quad 3\ldots18)$ represents the measured data of each verification point along the longitudinal direction of the box girder.

$$PD(\varepsilon) = \frac{\varepsilon_{V,i}^{iFEM} - \varepsilon_{V,i}^{VP}}{\varepsilon_{V,i}^{VP}} \tag{27}$$

The measured values near the neutral axis range from −6 to 44 με; the maximum measured strain value is found at the intersection of the top plate and the midship section, measuring 1216.6 $\mu\varepsilon$; and the minimum measured strain value is found at the intersection of the bottom plate and the midship section, measuring −1328.41 με. A maximum inaccuracy of −2563.51% exists between the measured value and the reconstructed value. The measured strain value is low since this site is close to the neutral axis. Due to the measuring point's inadequate layout precision, which results in a measured value of −15 με and a reconstructed value of 0.61 με, the measuring point may be slightly off from the neutral axis. The bottom surface displays a minimum error of 0.1% between the measured value and the reconstructed value. The relative errors between the measured value and the reconstructed value are all less than 10%, with the exception of the measuring spots close to the neutral axis.

Figure 12 vividly depicts the alignment trends between the measured values and reconstructed values for 45 points across three surfaces. Meanwhile, Table 3 provides the detailed data and relative errors for the points on these three panels.

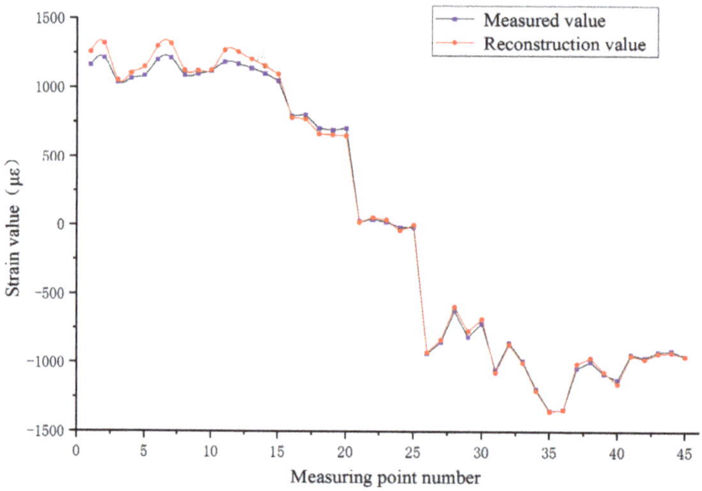


**Figure 12.** Strain trend of measured values and reconstructed values of box girder top surface–side surface–bottom surface.

**Table 3.** Table of relative errors of reconstruction points.

| Reconstruction Point Area | Reconstruction Point Number | Measured Value (με) | Reconstruction Value (με) | Relative Error |
|---|---|---|---|---|
| Top plate | 1 | 1163.68 | 1260.06 | −7.65% |
| | 2 | 1216.60 | 1320.45 | −7.86% |
| | 3 | 1039.00 | 1053.30 | −1.36% |
| | 4 | 1066.00 | 1103.55 | −3.40% |
| | 5 | 1087.42 | 1150.11 | −5.45% |
| | 6 | 1201.73 | 1301.33 | −7.65% |
| | 7 | 1215.69 | 1320.24 | −7.92% |
| | 8 | 1090.00 | 1123.39 | −2.97% |
| | 9 | 1098.00 | 1120.86 | −2.04% |
| | 10 | 1122.81 | 1125.56 | −0.24% |
| | 11 | 1185.75 | 1273.76 | −6.91% |
| | 12 | 1174.00 | 1263.91 | −7.11% |
| | 13 | 1143.00 | 1210.53 | −5.58% |
| | 14 | 1104.00 | 1159.82 | −4.81% |
| | 15 | 1052.05 | 1100.87 | −4.43% |

**Table 3.** *Cont.*

| Reconstruction Point Area | Reconstruction Point Number | Measured Value (με) | Reconstruction Value (με) | Relative Error |
|---|---|---|---|---|
| Port | 1 | 800.00 | 787.75 | 1.55% |
| | 2 | 807.00 | 776.86 | 3.88% |
| | 3 | 708.00 | 669.66 | 5.73% |
| | 4 | 696.00 | 662.56 | 5.05% |
| | 5 | 707.75 | 654.53 | 8.13% |
| | 6 | 27.00 | 23.15 | 16.62% |
| | 7 | 44.00 | 56.00 | −21.43% |
| | 8 | 24.00 | 38.96 | −38.39% |
| | 9 | −15.00 | −36.49 | −58.89% |
| | 10 | −15.00 | 0.61 | −2563.51% |
| | 11 | −933.00 | −926.05 | 0.75% |
| | 12 | −851.00 | −835.38 | 1.87% |
| | 13 | −620.00 | −590.98 | 4.91% |
| | 14 | −814.00 | −771.59 | 5.50% |
| | 15 | −712.29 | −678.83 | 4.93% |
| Bottom plate | 1 | −1053.00 | −1069.71 | 1.56% |
| | 2 | −854.00 | −865.98 | 1.38% |
| | 3 | −989.00 | −998.25 | 0.93% |
| | 4 | −1190.00 | −1201.75 | 0.98% |
| | 5 | −1348.98 | −1350.95 | 0.15% |
| | 6 | −1341.00 | −1340.94 | 0.00% |
| | 7 | −1040.00 | −1006.64 | −3.31% |
| | 8 | −992.00 | −963.82 | −2.92% |
| | 9 | −1078.00 | −1066.70 | −1.06% |
| | 10 | −1123.14 | −1152.12 | 2.52% |
| | 11 | −938.00 | −945.51 | 0.79% |
| | 12 | −962.00 | −971.23 | 0.95% |
| | 13 | −921.00 | −930.39 | 1.01% |
| | 14 | −911.00 | −923.32 | 1.33% |
| | 15 | −951.42 | −952.97 | 0.16% |

The side and close to the axis are where there is the greatest relative error in the rebuilt value of the verification point. Measuring points 6–10 on the port are arranged at the neutral axis position. Near the neutral axis of the box girder, the strain gradient becomes smaller, and the strain value gradually approaches zero. At the same time, the structure is easily affected by external disturbance, measurement accuracy, and material micro-inhomogeneity, which leads to the amplification of experimental measurement errors. The reconstruction process is influenced by iFEM algorithm, and the recognition degree of low strain region is low, which continues to expand the reconstruction error. Table 4 displays the relative errors of the other verification points, excluding the neutral axis, which ranges from −0.40% to 2.51%.

**Table 4.** Table of relative errors of verification points.

| Verification Point Area | Verification Point Number | Measured Value (με) | Reconstruction Value (με) | Relative Error |
|---|---|---|---|---|
| Top plate | 1 | 1104 | 1108.94 | −0.45% |
| | 2 | 940 | 939.15 | 0.09% |
| | 3 | 1107 | 1111.58 | −0.41% |
| | 4 | 1039 | 1036.08 | 0.28% |
| | 5 | 1128 | 1134.08 | −0.54% |
| | 6 | 1086 | 1090.37 | −0.40% |

**Table 4.** *Cont.*

| Verification Point Area | Verification Point Number | Measured Value (με) | Reconstruction Value (με) | Relative Error |
|---|---|---|---|---|
| Port | 1 | 712 | 703.05 | 1.27% |
| | 2 | 746 | 732.67 | 1.82% |
| | 3 | −13 | −10.76 | 20.81% |
| | 4 | −6 | 1.11 | −640.95% |
| | 5 | −530 | −517.04 | 2.51% |
| | 6 | −774 | −771.58 | 0.31% |
| Bottom plate | 1 | −1315 | −1328.41 | 1.01% |
| | 2 | −1166 | −1172.55 | 0.56% |
| | 3 | −1131 | −1138.52 | 0.66% |
| | 4 | −1114 | −1123.13 | 0.81% |
| | 5 | −861 | −856.05 | −0.58% |
| | 6 | −1116 | −1121.32 | 0.47% |

The number and position of sensors are very important to the accuracy of strain field reconstruction via the inverse finite element method. The more sensors there are, the higher the reconstruction accuracy is, but it also increases the calculation cost and prolongs the calculation time. The reasonable arrangement of sensor positions can significantly improve the reconstruction effect, especially in the stress concentration area, which can effectively capture key data and reduce the reconstruction error. Therefore, optimizing the number and location of sensors is the key to improving the accuracy of strain reconstruction.

### 5.2. Strain Field Reconstruction

The top plate's reconstructed value has a maximum relative error of 7.86% and a minimum relative error of 0.24%. According to Figure 13, the reconstructed value's trend and the measured value's trend are both in good agreement. Figures 14 and 15, respectively, depict the top surface's measured strain field and reconstructed strain field.

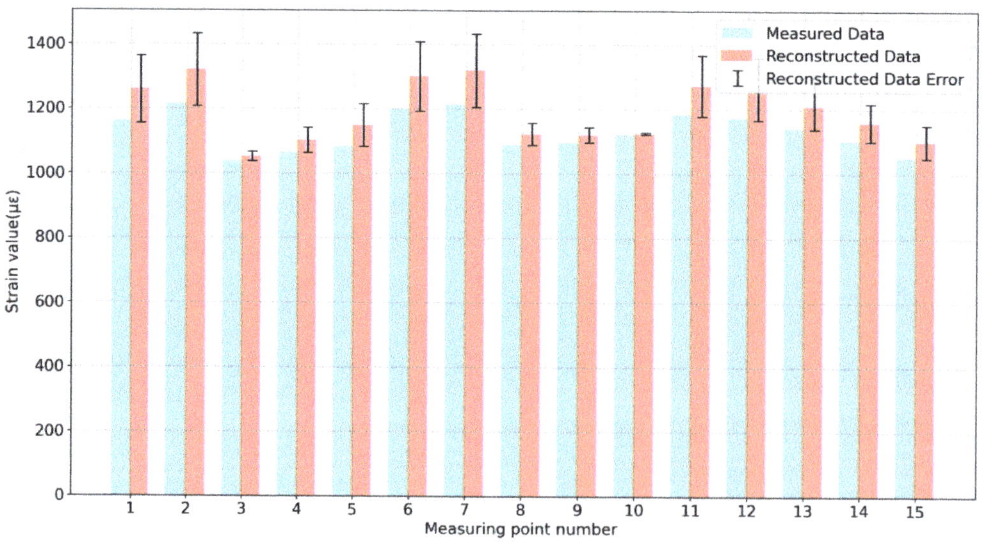


**Figure 13.** Comparison and error between measured values and reconstructed values of measuring points on the top surface.

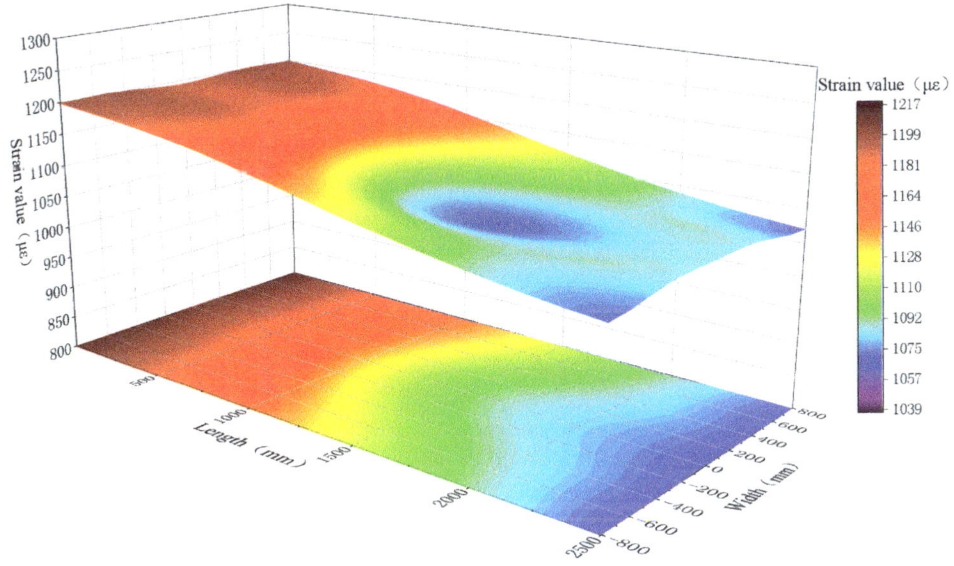


**Figure 14.** Measured strain field on top plate surface.

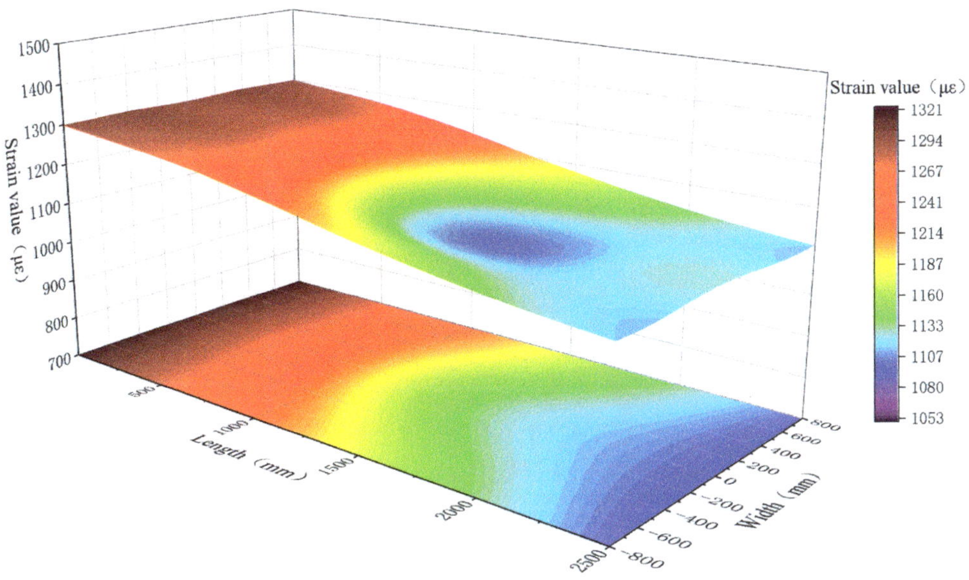


**Figure 15.** Reconstruction strain field of top plate surface.

In each strain field, the top one is drawn based on actual measurements or reconstructed data, while the bottom one represents the vertical projection of the top strain field within the plane.

Except for the errors of the measuring points close to the neutral axis on the side, the relative errors between the reconstructed values and the measured values of other measuring points are between 0.75% and 8.13%. Their trends are in good agreement, as shown in Figure 16, and the measured strain field and the reconstructed strain field on the broadside are shown in Figures 17 and 18, respectively.

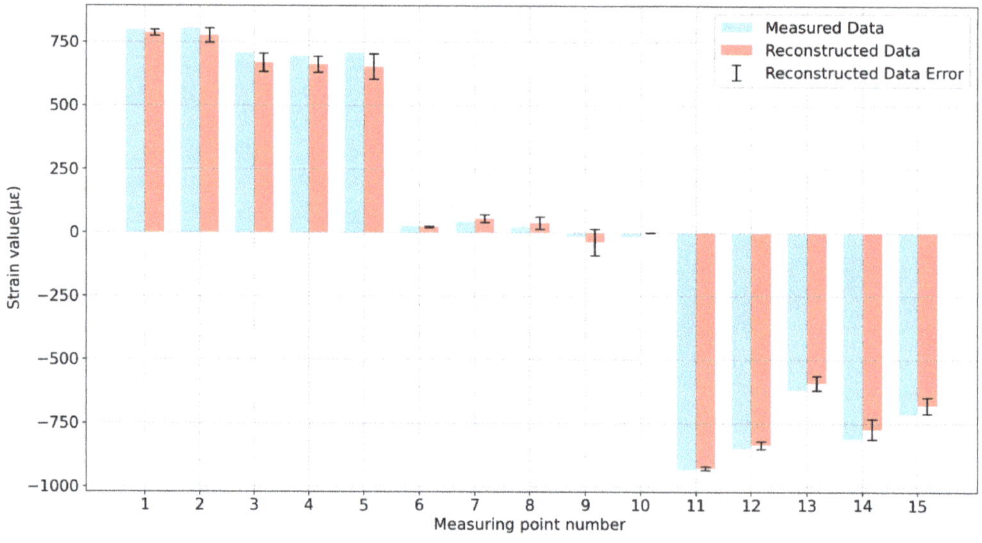


**Figure 16.** Comparison and error between measured values and reconstructed values of measuring points on the side.

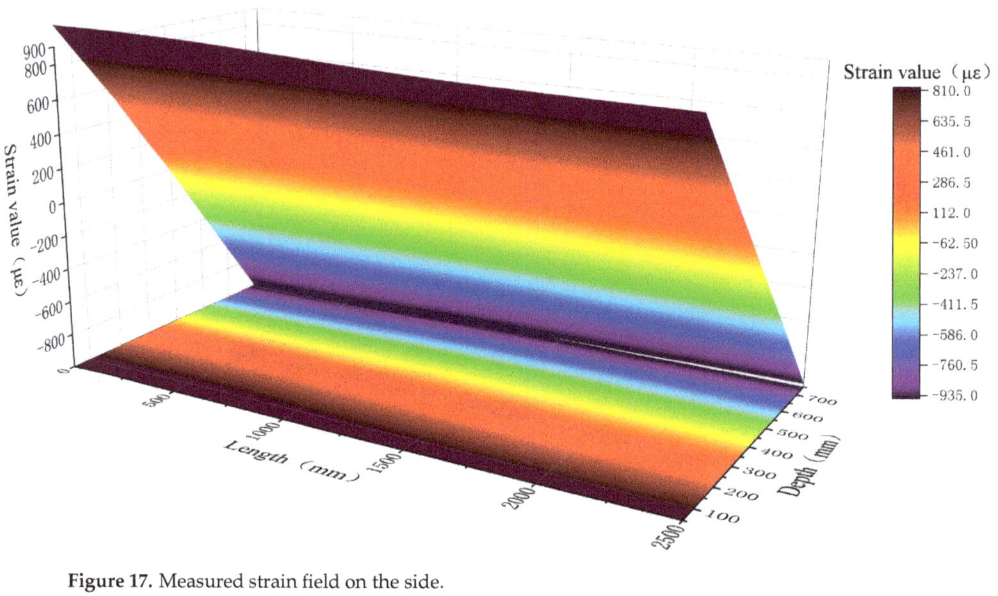


**Figure 17.** Measured strain field on the side.

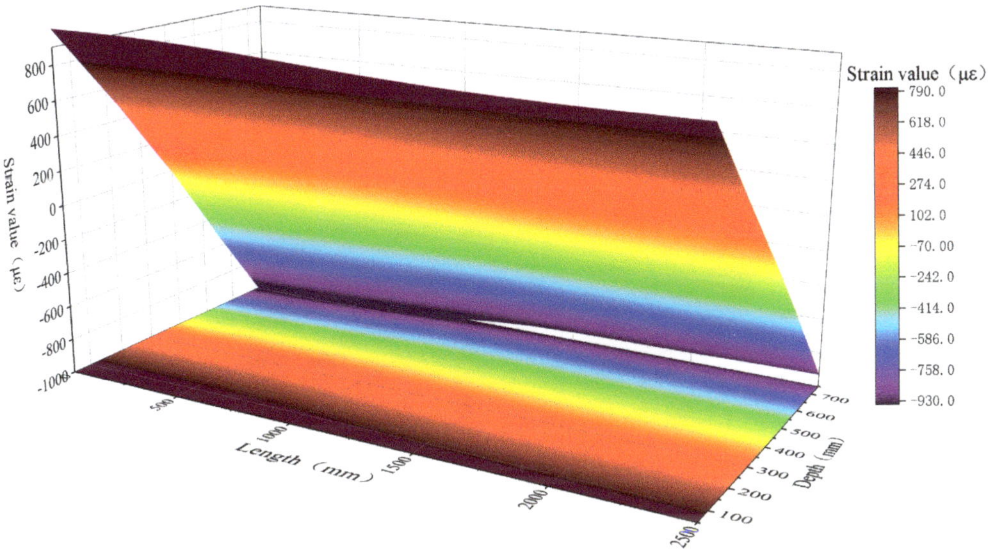


**Figure 18.** Strain field of the side reconstruction.

During the loading process, the box girder's bottom plate is compressed, which makes it vulnerable to widespread local deformation. Figure 19 illustrates the good agreement between the two trends and the relative error between the measured value and the reconstructed value of the measuring point, which ranges from −3.31% to 2.52%. Figures 20 and 21, respectively, depict the measured strain field and the reconstructed strain field of the ship's bottom plate.

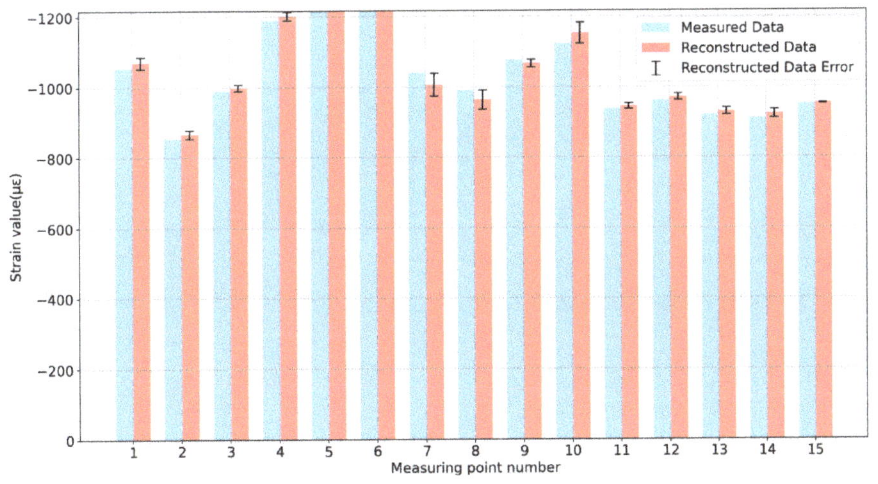


**Figure 19.** Comparison and error between measured values and reconstructed values of measuring points on the bottom surface.

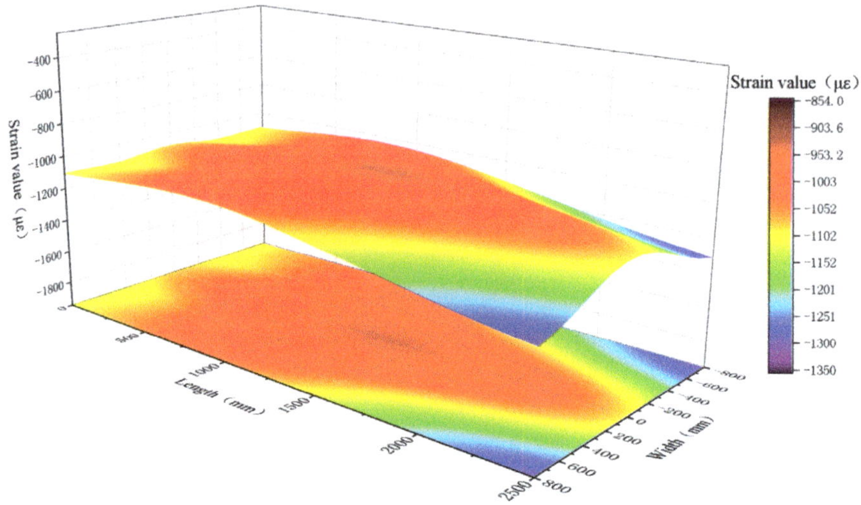


**Figure 20.** Measured strain field on bottom plate surface.

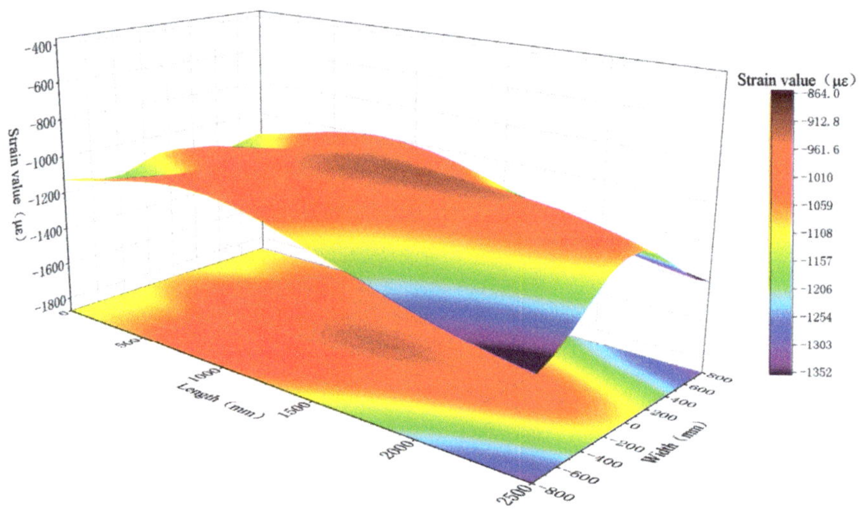


**Figure 21.** Reconstruction strain field of bottom plate surface.

## 6. Conclusions

This study uses iFEM technology and the iQS4 element to successfully recreate the strain field of a long-span box girder.

1. For the first time, significant quantities of measured data and virtual sensor strain data were used to reconstruct the strain field of the top plate, broadside, and bottom plate of the box girder test section under a certain working environment.
2. By confirming the reconstruction results, it was found that the average error between the reconstructed value and the measured value (again, excluding the measurement points near the neutral axis) was 3.42% and the reconstruction error of the verification points ranged from −0.40% to 2.51%.
3. In this study, a large number of experimental data were used to reconstruct the strain field of a long-span hull box girder and present it visually, allowing us to quickly

evaluate the stress state of the hull structure in complex sea conditions, prevent potential structural damage, and ensure the safety of ship structure.

In the future, iFEM will have broad prospects in the field of ship and ocean engineering health monitoring. The research directions that need to be further improved may include improving the rapidity and computational efficiency of the algorithm, optimizing the sensor layout strategy, and verifying its accuracy and reliability under more complex load conditions so as to further enhance the safety of ship structures and improve the life cycle.

**Author Contributions:** Methodology, G.C.; validation, P.W. and T.Z.; formal analysis, Z.J. and H.C.; investigation, N.Z.; data curation, F.L.; writing—original draft, G.C.; writing—review and editing, X.W.; supervision, X.W.; funding acquisition, X.W. All authors have read and agreed to the published version of the manuscript.

**Funding:** This paper was supported by program of China (WDZC70202010101), the National Key Research and Development Program of China (2021YFC2802300) and Intergovernmental International Science, Technology, and Innovation Cooperation Key Project under the National Key Research and Development Program of China (2024YFE0104300).

**Institutional Review Board Statement:** Not applicable.

**Informed Consent Statement:** Informed consent was obtained from all subjects involved in the study.

**Data Availability Statement:** Relevant research data are mostly displayed in the paper. If you need other data, you can contact the author by email (except private data).

**Acknowledgments:** Thanks to all the experimenters and participants in the research content of this paper.

**Conflicts of Interest:** The authors declare no conflicts of interest.

## References


1. Wang, X.; Zhu, Q.; Zhang, T.; Yang, H.; Chen, G. Research progress of ship structure monitoring technology. *J. Ship Mech.* **2022**, *26*, 1246–1253.
2. Zheng, Q.; Wang, X.; Zhao, X.; Yang, H.; Xu, C.; Jiang, Z. Research on strain sensor technology applied to structural safety monitoring of ships and offshore engineering. *Equip. Environ. Eng.* **2023**, *20*, 83–89.
3. Bernasconi, O.; Ewins, D.J. Application of strain modal testing to real structures. In Proceedings of the 7th International Modal Analysis Conference, Las Vegas, NV, USA, 30 January–2 February 1989; Society for Experimental Mechanics: Bethel, CT, USA, 1989; Volume 2, pp. 1453–1464.
4. Chopra, I. Review of state of art of smart structures and integrated systems. *AIAA J.* **2002**, *40*, 2145–2187. [CrossRef]
5. Sun, J. Inverstigation on Morphing Wing Structures Based on Shape Memory Polymer Composite (SMPC) Skins and Active Honeycomb Structures. Ph.D. Thesis, Harbin Institute of Technology, Harbin, China, 2015.
6. Ko, W.L.; Richards, W.L.; Tran, V.T. Displacement Theories for In-Flight Deformed Shape Predictions of Aerospace Structures. NASA/TP-2007-214612, 1 October 2007.
7. Derkevorkian, A.; Masri, S.F.; Alvarenga, J.; Boussalis, H.; Bakalyar, J.; Richards, W.L. Strain-based deformation shape-estimation algorithm for control and monitoring applications. *AIAA J.* **2013**, *51*, 2231–2240. [CrossRef]
8. Zhang, T.; Hu, J.; Wang, X.; Chen, G.; Zhu, Q.; Jiang, Z.; Wang, Z. Solving approach for global stress field of the 3D structures based on artificial intelligence. *J. Ship Mech.* **2023**, *27*, 238–249.
9. Kefal, A.; Oterkus, E. Displacement and stress monitoring of a Panamax containership using inverse finite element method. *Ocean. Eng.* **2016**, *119*, 16–29. [CrossRef]
10. Kefal, A. An efficient curved inverse-shell element for shape sensing and structural health monitoring of cylindrical marine structures. *Ocean. Eng.* **2019**, *188*, 106262. [CrossRef]
11. Huang, H. Research on Deformation Reconstruction of Plateand Girder Structure Based on Inverse Finite Element Method. Master's Thesis, University of Science and Technology, Huazhong, China, 2021. [CrossRef]
12. Mao, Z. Comparison of Shape Reconstruction Strategies in a Complex Flexible Structure. Master's Thesis, University of California, San Diego, CA, USA, 2008.
13. Tessler, A.; Spangler, J. *A Variational Principle for Reconstruction of Elastic Deformations in Shear Deformable Plates and Shells*; National Aeronautics and Space Administration, Langley Research Center: Hampton, VA, USA, 2003.
14. Hu, M.; Wu, S.; Dong, E. Deformation reconstruction of a three-dimensional panel structure opon IFEM. *Eng. Mech.* **2024**, *41*, 235–245.
15. Ke, Z.; Yuan, S.; Ren, Y.; Xu, Y. Shape reconstruction of self-adaptive morphing wings' fishbone based on iFEM. *Acta Aeronaut.* **2020**, *41*, 250–260.

16. Kefal, A.; Oterkus, E. Structural health monitoring of marine structures by using inverse finite element method. In *Analysis and Design of Marine Structures V*, 1st ed.; CRC Press: Boca Raton, FL, USA, 2015; pp. 341–349.
17. Kefal, A.; Oterkus, E. Displacement and stress monitoring of a chemical tanker based on inverse finite element method. *Ocean. Eng.* **2016**, *112*, 33–46. [CrossRef]
18. Kefal, A.; Mayang, J.B.; Oterkus, E.; Yildiz, M. Three dimensional shape and stress monitoring of bulk carriers based on iFEM methodology. *Ocean. Eng.* **2018**, *147*, 256–267. [CrossRef]
19. Putranto, T.; Kõrgesaar, M.; Tabri, K. Application of equivalent single layer approach for ultimate strength analyses of ship hull girder. *J. Mar. Sci. Eng.* **2022**, *10*, 1530. [CrossRef]
20. Putranto, T. Equivalent single layer approach for ultimate strength analysis of box girder under bending load. *Ocean. Eng.* **2024**, *292*, 116535. [CrossRef]
21. Yan, H.; Bai, C.; Jia, T.; Liu, C. Application of inverse finite element in deformation and reconstruction of ship reinforced plate structure. *Shipbuild. China* **2023**, *64*, 168–179.
22. Zhu, Q.; Wu, G.; Zeng, J.; Jiang, Z.; Yue, Y.; Xiang, C.; Zhan, J.; Zhao, B. Enhanced Strain Field Reconstruction in Ship Stiffened Panels Using Optical Fiber Sensors and the Strain Function-Inverse Finite Element Method. *Appl. Sci.* **2023**, *14*, 370. [CrossRef]
23. Wei, P.; Li, C.; Jiang, Z.; Wang, D. Real-Time Digital Twin of Ship Structure Deformation Field Based on the Inverse Finite Element Method. *J. Mar. Sci. Eng.* **2024**, *12*, 257. [CrossRef]
24. Vazquez, S.L.; Tessler, A.; Quach, C.C.; Cooper, E.G.; Parks, J.; Spangler, J.L. Structural Health Monitoring Using High-Density Fiber Optic Strain Sensor and Inverse Finite Element Methods. NASA/TM-2005-213761, 1 May 2005.
25. Tessler, A.; Spangler, J.L. Inverse FEM for full-field reconstruction of elastic deformations in shear deformable plates and shells. In Proceedings of the 2nd European Workshop on Structural Health Monitoring, Munich, Germany, 7–9 July 2004.
26. Kefal, A.; Oterkus, E.; Tessler, A.; Spangler, J.L. A quadrilateral inverse-shell element with drilling degrees of freedom for shape sensing and structural health monitoring. *Eng. Sci. Technol. Int. J.* **2016**, *19*, 1299–1313. [CrossRef]
27. Cao, Z.; Fei, Q.; Jiang, D.; Zhang, D.; Jin, H.; Zhu, R. Dynamic sensitivity-based finite element model updating for nonlinear structures using time-domain responses. *Int. J. Mech. Sci.* **2020**, *184*, 105788. [CrossRef]

Journal of
*Marine Science and Engineering*

MDPI

*Article*

# A Virtual Assembly Technology for Virtual–Real Fusion Interaction of Ship Structure Based on Three-Level Collision Detection

**Ze Jiang [1,2], Pengyu Wei [1,2,*], Yuntong Du [1], Jiayi Peng [1] and Qingbo Zeng [1,2]**

[1] China Ship Scientific Research Centre, Wuxi 214082, China; jzcssrc@163.com (Z.J.); yuntong1992@163.com (Y.D.); 15961879552@163.com (J.P.); zengqingbo1986@163.com (Q.Z.)
[2] National Key Laboratory of Ship Structural Safety, Wuxi 214082, China
* Correspondence: wei_pengyu@163.com

**Abstract:** With the rapid advancement of new-generation information technology, the virtual–real fusion interaction has increasingly become a crucial technique for structural analysis to determine the strength envelope of hulls. A high-precision assembly of the experimental devices in a virtual environment is vital. A virtual assembly method for a structural virtual–real fusion test based on the oriented bounding box (OBB) algorithm, the Devillers and Guigue algorithm, and differential triangle facets algorithm are proposed in this paper. The experiment of the connector in a typical offshore floating platform is performed as a case, which indicates that the virtual assembly method proposed in this paper enables the assessment of assembly virtually prior to the actual experiment, and the assembly accuracy can reach 0.01 mm. The digitalization and virtual–real fusion interaction for the mechanical experiment of hulls are advanced to ensure efficiency, safety, and economy.

**Keywords:** virtual assembly; virtual–real fusion interaction; ship structure; collision detection

## 1. Introduction

Physical and virtual experiments are the two main approaches used in the field of ship structural performance assessment to determine structural performance indicators [1,2]. Although physical experiments have continuously evolved, issues such as high costs, long cycles, and scale effects still exist. Therefore, virtual experiments must be conducted to analyze the failure behavior of the structure. In recent years, virtual experiments have been increasingly used as an addition to physical experiments, but issues of varied calculation outputs and reliability remain obstacles. Ship performance must be improved since ships are facing more complex and diverse environmental loads like wind, waves, and currents. Simultaneously, the creative development of ship structures and rapid product iterations are needed. The current research on the fusion of physical and virtual experiments of ship structures struggles to meet demands. This is mainly manifested in the fact that physical experiment data have real-time capabilities, but most of them are discrete point data measured by sensors, which cannot reflect the information of the entire structure. On the contrary, although virtual test data can obtain full field information of the structure, these data are generally calculated using CAE software (abaqus6.14) before the test, and the real-time calculation efficiency during the test process still cannot meet practical requirements. The virtual–real fusion experiments [3] for ship structures, based on high-fidelity virtual models, reproduce the physical test system in a virtual space using high-precision virtual assembly techniques. This approach establishes a digital link between the physical entity and the virtual model during experiments by integrating virtual–real data analysis, synchronous mapping algorithms, and comprehensive visualization technologies, representing a novel direction in the field of ship structure experiments. Virtual assembly technology, as a key step in virtual–real fusion experiments for ship structures, is worthy of further study.

Virtual assembly technology originated in the early 1980s. The design and manufacture of large structures such as aircraft and ships have been revolutionized with the development of virtual assembly techniques [4]. The primary goal of virtual assembly technology is to generate a digital model that can substitute for the physical prototype traditionally used in structural research and development. This allows for the visualization of the assembly process design, simulation, and quality evaluation within a virtual space. Over several decades, virtual assembly technology has evolved into a comprehensive assembly process planning technique that integrates virtual reality, mechanical modeling, computer networks, and artificial intelligence [5]. Assembly sequence [6–9], assembly path [10–12], collision detection, and human factors engineering [13,14] have been focused on.

In the virtual–real fusion experiments of ship structures, virtual assembly technology aims to conduct the assembly design and physical scene presentation of the experimental loading devices, connecting devices, and experimental objects within a virtual environment. This technology also enables the virtual mapping description and visualization of the physical test assembly process. The installation feasibility of each component in the physical test design scheme is verified through the virtual assembly. Additionally, interference inspection of potential collisions and movements during the experimental process is conducted to ensure that the designs of experiments achieve the desired outcome. Consequently, collision detection is the core technology of the virtual assembly in the virtual–real fusion experiments of ship structures. Current collision detection methods primarily include the spatial decomposition method [15,16] and the hierarchical bounding box method [17–19]. Based on these methods, numerous researchers have proposed corresponding collision detection algorithms for different detection objects, typically divided into rough detection and detailed detection. For rough detection, the collision volume bounding box algorithm is primarily utilized, such as the bounding sphere [20], AABB (Axis-Aligned Bounding Box) [21], OBB (Oriented Bounding Box) [22], k-DOP (K-Discrete orientation polytope) bounding box [23], and mixed bounding box [24]. These algorithms only need to determine if there is an intersection between geometries, making them efficient but not precise. For detailed detection, the triangular facets algorithm is primarily utilized. Kim [25] proposed a hierarchical collision detection algorithm that combines CPU and GPU usage. The CPU is employed for traversal and coarse detection of hierarchical bounding bodies, while the GPU handles basic collision detection of triangular facets. Pabst [26] uses the CPU to update the bounding volume and triangular facets and performs both rough and detailed collision detection on the GPU. Through this task decomposition strategy, the algorithm can efficiently use a uniform grid to divide space, thereby maximizing the speed of the collision detection process. Therefore, the triangular facets algorithm is more suitable for interference detection of complex contour models, offering significantly higher detection accuracy than the collision volume bounding box algorithm, though at the cost of reduced detection efficiency.

In conclusion, the applications of virtual assembly technology have been extended to various fields including advanced equipment manufacturing, product design, training, and maintenance. He et al. proposed an accurate 3D reconstruction method for ship component models to assemble ship rib models with manufacturing errors, reducing interference in ship rib assembly by 50% and increasing efficiency by 24% [27]. Lau pointed out that simulation technologies for simulation, visualization, and modeling, such as augmented reality and virtual reality (AR, VR), are rapidly developing. A series of new modeling and simulation technologies, such as augmented reality, virtual reality, virtual assembly, and the use of 3D ship models, are reshaping the methods of ship design, safe operation, and training [28]. Concurrently, advancements have been made in the associated collision detection algorithms. Nevertheless, literature addressing the applications of virtual assembly technology in ship structure testing remains scarce. Presently, the assembly of physical systems for ship structure tests predominantly relies on personal experience, with an accuracy of no more than 5 mm. Therefore, the success rate of one-time assembly is low, which leads to the need for polishing or even reprocessing of the structure (from one day to one week), seriously affecting the normal conduct of the test and violating the requirements of refined development of the test. This paper introduces a high-precision virtual assembly technique tailored for the virtual–real fusion experiments of ship structures. This method employs a hierarchical combination of the OBB algorithm, the Devillers and Guigue algorithm, and the differential triangle facets algorithm to facilitate both qualitative and quantitative assessments of three-level collision detection during the virtual model assembly process. The assembly procedure is subsequently visualized in a virtual environment using the Unity3D platform, thereby enhancing on-site assembly efficiency and reducing experimental expenses. Following the introduction, the principles of the OBB algorithm, the Devillers Guigue algorithm, the differential triangle facets algorithm, and the virtual assembly method utilizing a three-level collision detection are presented in Section 2. In Section 3, the efficacy and dependability of the proposed method are validated through an experiment on a typical offshore platform connector structure, which demonstrates the virtual assembly process of a structure for a virtual–real fusion experiment. Conclusions are drawn in Section 4.

## 2. Methods

### 2.1. Virtual Assembly Method Based on Three-Level Collision Detection

The high-precision virtual assembly method is proposed based on the Unity3D software platform (unity2019.4.23f1c1), which enables the simulation and quantitative inspection of the assembly process through a three-level collision detection. As illustrated in Figure 1, the method consists of a primary interference check using an OBB algorithm, an intermediate interference check using the Devillers and Guigue algorithm, and an advanced interference check using a differential triangle facets algorithm. The reason for choosing these three methods is that compared to AABB and bounding ball algorithms, the OBB algorithm can better fit the shape of objects while ensuring efficiency. The Devillers and Guigue algorithm and differential triangle facets algorithm are choices that fully consider assembly accuracy and sacrifice some computational efficiency.

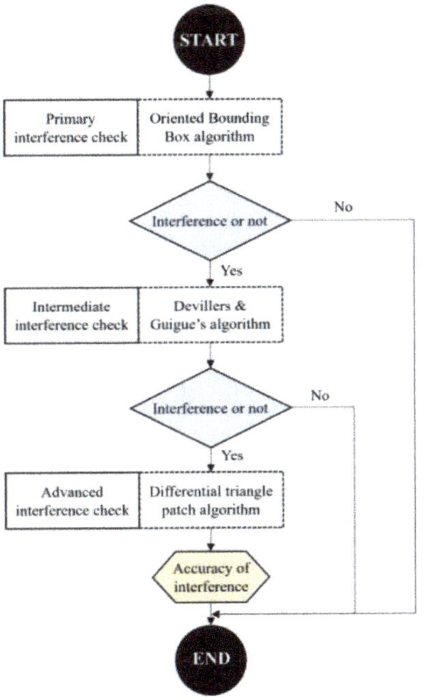


**Figure 1.** Three-level collision detection workflow diagram.

By using the OBB algorithm, a rough interference check is conducted as the primary interference check on the object to be tested and the object that may collide is retained. Based on the separating axis theory, the bounding box of the object to be measured is projected onto different separating axes, and their projections are compared. If the projections intersect, interference will occur. When a potential interference within the bounding box is detected, the intermediate interference check is initiated instead. The Devillers and Guigue algorithm used in the intermediate interference check is a typical vector discriminant algorithm that can determine the positional relationship between two triangles by a series of calculated values, thereby determining the intersection of two triangular facets. This algorithm is an improvement and optimization of the scalar discrimination algorithm, which enhances the efficiency and stability of the algorithm. At this stage, although whether the virtual model interferes can be determined, the scale of interference cannot be accurately measured. Consequently, the differential triangle facets algorithm is used as the advanced interference inspection to further differentiate the virtual model's triangular facets into quadrilateral facets that meet the design accuracy requirements. The specific size parameter information of the virtual model interference is obtained in this step, and the complex physical experiments in virtual assembly design functions are hence realized.

### 2.2. Collision Algorithm

#### 2.2.1. Initial Interference Check Based on OBB Bounding Box Algorithm

The OBB algorithm is a method that involves creating a minimal cuboid surrounding geometric objects, with any arbitrary orientation within the coordinate system. The key to constructing an OBB lies in identifying the optimal orientation of the box and ascertaining the minimum dimensions of its length, width, and height. The detailed construction process is below.

Suppose that G is a geometric entity. The aggregate of all elementary geometric components, such as triangles and circles, is represented as $S_G$, presuming that the elementary constituents of the object are exclusively triangles and denoting their total count as n, wherein the three vertices of the i-th triangle are labeled as $a_i$, $b_i$, and $c_i$. The subsequent steps detail the evaluation procedures for the mean $\mu$ and covariance C of the set $S_G$.

$$\mu = \frac{1}{3n}\sum_{i=1}^{n}(a_i + b_i + c_i) \tag{1}$$

$$C_{jk} = \frac{1}{3n}\sum_{i=1}^{n}(a_j^i \cdot a_k^i + b_j^i \cdot b_k^i + c_j^i \cdot c_k^i) \tag{2}$$

where $C_{jk}$ represents a numerical value, and the matrix with elements $C_{jk}$ is denoted as C. Given that the derived matrix C is a real symmetric matrix possessing three distinct eigenvalues, the eigenvectors associated with these three eigenvalues are orthogonal to one another. These eigenvectors, once normalized, yield a basis set that forms the three directional axes of the OBB, thereby establishing the optimal orientation of the OBB. Next, the minimum length, width, and height of OBB are determined. To pinpoint these minima, it suffices to project the vertices of each elementary geometric entity in SG onto the aforementioned directional axes. The minimal value derived from each projection corresponds to the requisite dimensions of length, width, and height, thus finalizing the construction of the OBB.

As depicted in Figure 2, the OBB for the virtual prototype of the test article is erected in Unity3D, utilizing the previously elucidated algorithm. A1 and A2 represent disparate test objects, while B symbolizes the loading device. Through the primary interference check algorithmic system, it is ascertained that B does not interfere with A1 but exhibits interference with A2. This observation underscores the potential for misjudgment in the primary interference check, thereby necessitating a more detailed examination of triangular patch intersections.

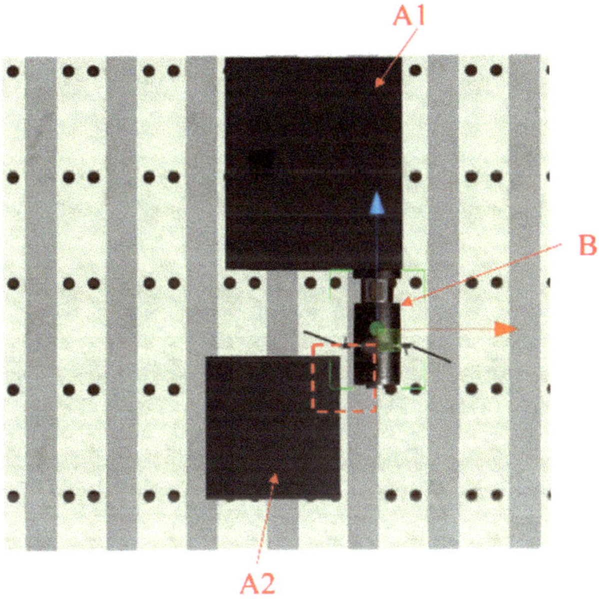


**Figure 2.** Primary interference check based on OBB bounding box algorithm.

### 2.2.2. Intermediate Interference Check Based on Devillers and Guigue Algorithm

The relative positional relationships among points, lines, and planes are assessed by evaluating the positive and negative determinants that correspond to the coordinates of each triangle's vertices according to the algorithm. Model interference occurs when two triangles intersect. The fundamental principle of this method is as follows:

Consider four points, $a$, $b$, $c$, and $d$, located in the space:

$$\begin{array}{l} a = (a_x, a_y, a_z) \\ b = (b_x, b_y, b_z) \\ c = (c_x, c_y, c_z) \\ d = (d_x, d_y, d_z) \end{array} \tag{3}$$

Define the determinant:

$$[a,b,c,d] = \begin{vmatrix} a_x & a_y & a_z & 1 \\ b_x & b_y & b_z & 1 \\ c_x & c_y & c_z & 1 \\ d_x & d_y & d_z & 1 \end{vmatrix} = \begin{vmatrix} a_x - d_x & a_y - d_y & a_z - d_z \\ b_x - d_x & b_y - d_y & b_z - d_z \\ c_x - d_x & c_y - d_y & c_z - d_z \end{vmatrix} \tag{4}$$

Utilizing the right-hand spiral rule, the spatial arrangement of four points can be ascertained by examining the sign of the determinant formed by the matrix $[a,b,c,d]$. For the triangle $\Delta abc$ configured in a counterclockwise sequence, if the determinant $[a,b,c,d]$ is positive, point $d$ resides in the direction normal to the plane of $\Delta abc$ (i.e., above the plane). Conversely, if the determinant is negative, $d$ lies in the opposite direction of the positive normal to the plane of $\Delta abc$ (i.e., beneath the plane). When the determinant equals zero, point d lies within the plane defined by $\Delta abc$.

Based on the aforementioned theory, consider two triangles, $T_1$ and $T_2$. Let $M_1^1, M_2^1, M_3^1$ be the vertices of triangle $T_1$, and let $M_1^2, M_2^2, M_3^2$ be the vertices of triangle $T_2$. The planes containing $T_1$ and $T_2$ are denoted by $\pi 1$ and $\pi 2$, respectively, as illustrated in Figure 3.

In the plane where two triangles $T_1$ and $T_2$ are situated, initially, non-intersection of $T_1$ and $T_2$ is excluded by assessing their positional relationship. According to the aforementioned theory, the positional relationship between $T_1$ and $T_2$ can be determined by the sign of the determinants formed by vectors $[M_1^2, M_2^2, M_3^2, M_i^1](i = 1, 2, 3)$. It is only when all corresponding determinants are non-zero and share the same sign that one can conclude that $T_1$ and $T_2$ do not intersect. Otherwise, intersection occurs, which can be categorized into three scenarios:

1. If all determinant values are zero, then $T_1$ and $T_2$ are coplanar.
2. If one determinant value is zero while the other two have identical signs, and if the test vertex lies inside $T_2$, an intersection occurs.
3. In other cases, the vertices of $T_1$ lie on opposite sides of the plane $\pi 2$ (including instances where an edge of $T_1$ lies within the plane $\pi 2$).

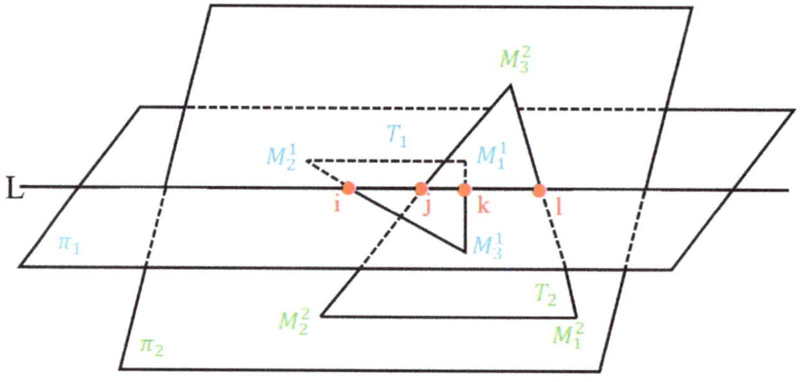


**Figure 3.** Schematic diagram of intersection of Devillers and Guigue algorithm.

For each triangle's vertices, a cyclic replacement operation is performed so that vertex $M_1^1(M_1^2)$ is situated on one side of the plane $\pi 2$ ($\pi 1$), while the other two vertices are on the opposite side. For vertices $M_2^2$ and $M_3^2(M_2^1, M_3^1)$, simultaneous exchange operations ensure that vertex $M_1^1(M_1^2)$ is positioned above the plane $\pi 2$ ($\pi 1$) in the direction normal to it. Through these preliminary exclusions and substitutions, adjacent edges of vertices $M_2^1M_3^1$, $M_3^1M_1^1$, and $M_2^2M_3^2$, $M_3^2M_1^2$ intersect at points i, j, k, l (where $i < j < k < l$). These four points correspond to two closed intervals on L: L1 = [i, j] and L2 = [k, l]. Consequently, the problem of testing the intersection for two triangles can be transformed into assessing the overlap between two closed intervals, L1 and L2. When L1 and L2 overlap, the triangles intersect; otherwise, they do not intersect.

In summary, because each triangle is a subset of its respective plane, when two triangles intersect, they must intersect along the intersection line of their respective planes. The process of interference inspection using the Devillers and Guigue algorithm is illustrated in Figure 4. This inspection method initially excludes non-intersecting triangles based on coplanarity, thereby significantly reducing the computational cost of testing numerous triangle pairs. By assessing the determinant's sign associated with each triangle vertex, the relative positional relationship between the two triangles is determined. This approach not only reduces computational costs but also enhances efficiency and stability in the calculation process. Consequently, it facilitates the qualitative judgment of virtual assembly interference detection, as depicted in Figure 5.

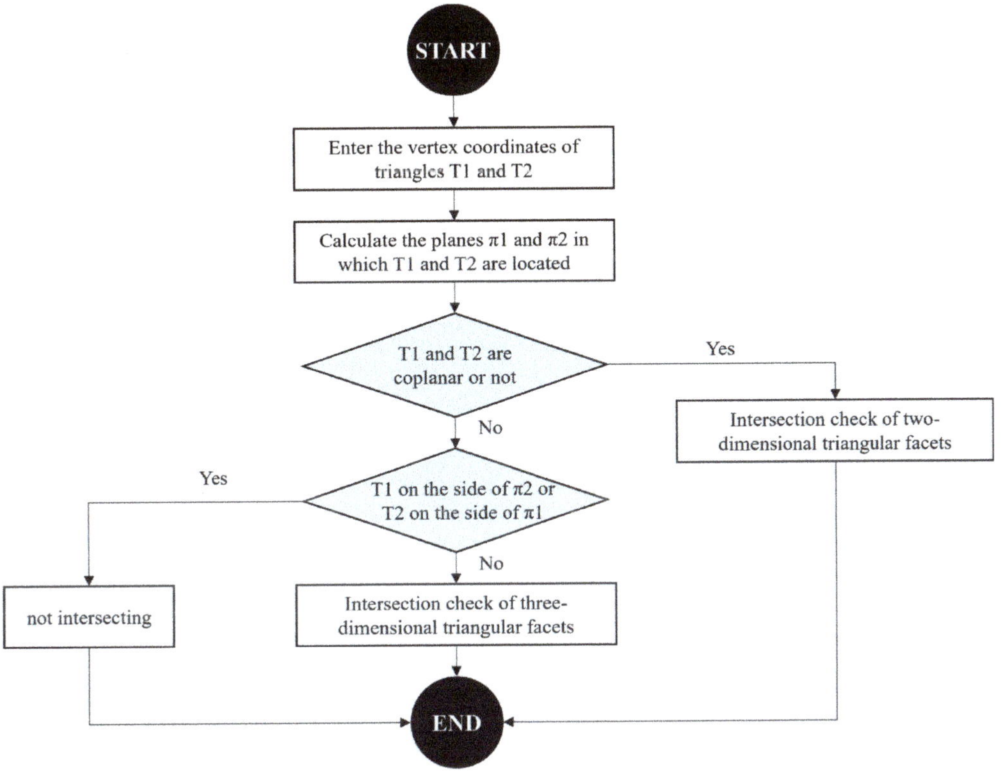


**Figure 4.** Flowchart of intermediate interference check based on Devillers and Guigue algorithm.

**Figure 5.** Schematic diagram of the intermediate interference check.

### 2.2.3. Advanced Interference Check Based on Differential Triangle Facets Algorithm

After conducting two levels of checks, the system can accurately detect interference between virtual models. However, it still lacks the ability to provide specific parameter information about model interference, which is crucial for quantitatively guiding the assembly of physical test systems. An advanced interference check method based on a differential triangle facets algorithm is thus presented. The triangular facet is divided into smaller square facets, which are then used for interference detection. As illustrated in Figure 6, taking a simple triangular-faceted slice as an example, the AB edges are segmented at equal intervals with a segmentation length set to P (where P is a constant). n points are then inserted into the AB edges, labeled as $f_1h_0, f_2h_0, f_3h_0, \ldots, f_nh_0$ according to Equation (5). The remaining values of R have minimal impact on the algorithm's efficacy and can thus be disregarded. Once the coordinates of points A and B are established, the positions of points $f_1h_0, f_2h_0, f_3h_0, \ldots, f_nh_0$ can be determined using vectors as shown in Equation (6).

$$R = L_{AB} \bmod P \tag{5}$$

$$f_nh_0(x_{f_nh_0}, y_{f_nh_0}, z_{f_nh_0}) = n \times P \times \frac{\vec{AB}}{|AB|} + A(x_A, y_A, z_A) \tag{6}$$

where $x_{f_nh_0}, y_{f_nh_0}, z_{f_nh_0}$ represent the 3D coordinates of the point $f_nh_0$, and $x_A, y_A, z_A$ denote the 3D coordinates of vertex A. Subsequently, the parallel lines that pass through points $f_1h_0$, $f_2h_0$, $f_3h_0$, ……, $f_nh_0$ and are parallel to the edge BC. These lines intersect with edge AC at points $Q_1$, $Q_2$, $Q_3$, …, $Qn$, respectively. Similarly, the lengths $P$ to $K_1Q_1$, $K_2Q_2$, $K_3Q_3$, …, $K_nQ_n$ on each side of the equidistant differential are determined. The coordinates of each partition point are then calculated. The triangle similarity property is employed to solve for the lengths of sides $K_1Q_1$, $K_2Q_2$, $K_3Q_3$, …, $K_nQ_n$, as demonstrated in Equation (7). This is followed by the equidistant differentiation of each side according to Equation (8), after which the coordinates of each cut-off point are determined using Equation (9).

$$\frac{L_{K_nQ_n}}{L_{BC}} = \frac{L_{AK_n}}{L_{AB}} = \frac{n \times P}{L_{AB}} \tag{7}$$

$$R = L_{K_nQ_n} \bmod P \tag{8}$$

$$f_nh_m(x_{f_nh_m}, y_{f_nh_m}, z_{f_nh_m}) = n \times P \times \frac{\vec{BC}}{|BC|} + f_nh_0(x_{f_nh_0}, y_{f_nh_0}, z_{f_nh_0}) \tag{9}$$

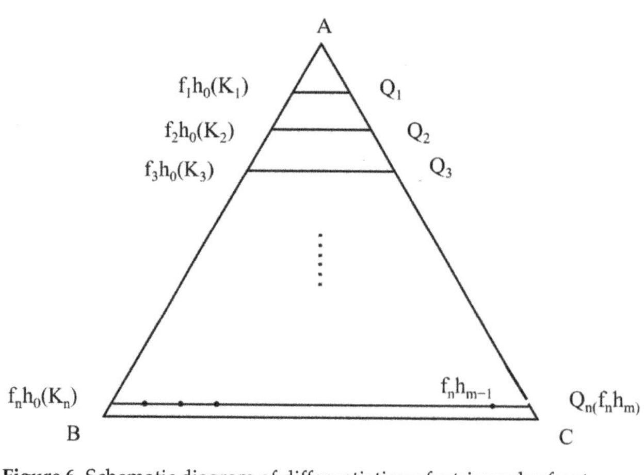


**Figure 6.** Schematic diagram of differentiation of a triangular facet.

Ultimately, the initial triangular facet is substituted by a quadrilateral facet with side length $P$, which is created using all the division points of the triangular facet as central points. Similarly, by applying this operation to all the triangular facets that constitute the model, the original triangular facets can be effectively replaced by the newly generated quadrilateral facets. The specific execution process of interference checking based on the differential triangle facets algorithm is illustrated, as shown in Figure 7. Building upon intermediate interference checking, further subdivision of the triangular facets that make up the virtual model allows for accurate acquisition of the interference dimensions of the two models by controlling the segmentation length P. This confirms the feasibility of the test assembly scheme.

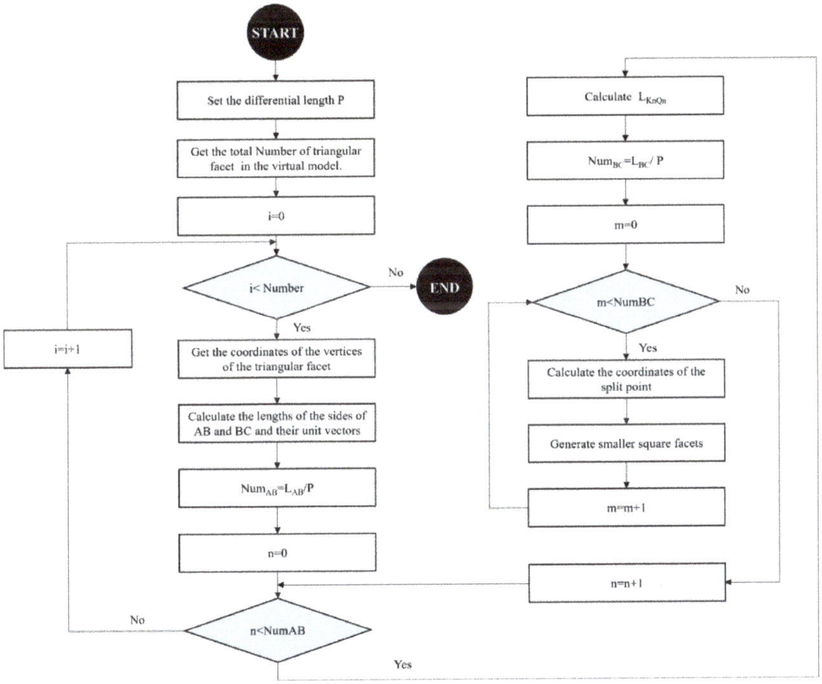


**Figure 7.** Flowchart of advanced interference check based on a differential triangle facets algorithm.

## 3. Application of High-Precision Virtual Assembly Method in Model Testing of Typical Offshore Platform Connectors

As an essential tool for ocean exploration, conservation, and exploitation, it is critical to guarantee the structural integrity of the offshore platform in the hostile maritime environment. The articulated connector used in this offshore platform is shown in Figure 8. To ascertain the safety and reliability of the articulated connector under extreme marine conditions, a scaled-down model of the connector at a ratio of 1:2.5 was designed and constructed for structural safety assessments. The test prototype and scaled-down testing apparatus are illustrated in Figure 9. This system comprises over ten components, including the bow connector, stern connector, connecting pins and bearings, cylinder actuators, reaction frames, and jigs. The installation procedure is intricate. It should be noted that the bow–stern connector is affixed by bearings and pins. A gap that is too small might prevent installation and the test model would need to be reprocessed. Conversely, an excessively large gap may make it difficult to meet the goals of the model test. Consequently, the virtual assembly technology is used to propose accurate machining accuracy requirements for the pin shaft, and the assembly scheme of the system is verified to ensure the smooth progress of the experiment. The high-fidelity virtual models are constructed in the structural virtual reality fusion testing platform, as shown in Figure 10.

**Figure 8.** Floating platform connector.

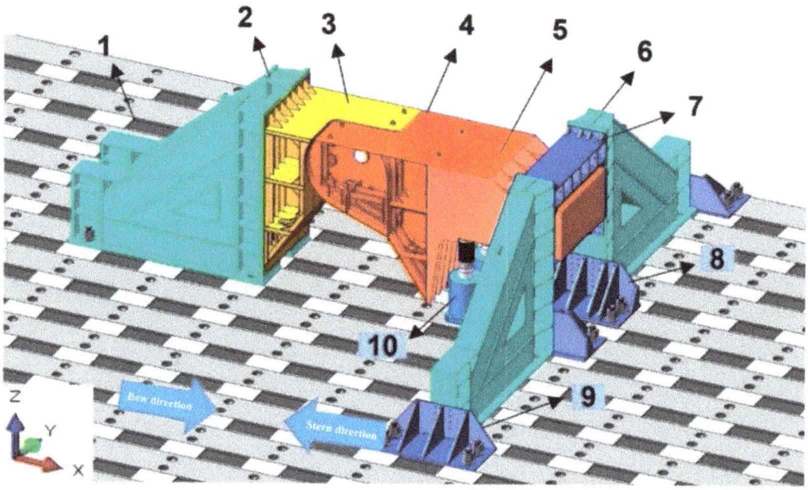


**Figure 9.** Schematic diagram of the connector model testing system (1—test platform, 2—reaction frame, 3—bow connector, 4—stern connector, 5—loading device, 6—transverse reaction frame, 7—beam, 8—inner fixing fixtures, 9—outer fixing fixtures, 10—cylinder actuator).

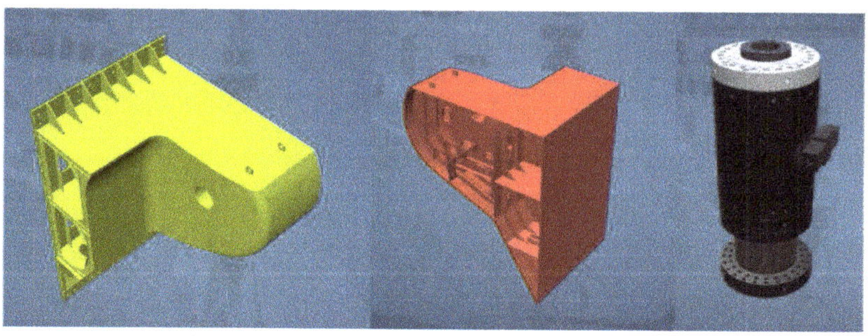

(a) (b) (c)

**Figure 10.** High-fidelity virtual modeling of the connector model test system. (**a**) Virtual model of bow connector, (**b**) virtual model of stern connector, and (**c**) virtual model of cylinder actuator.

Four reaction frames, a beam, and two sets of inner and outer fixed fixtures are used to establish the foundational framework for the testing system. These models are all reused in the laboratory, so laser scanners are used to obtain their STL models and import them into the system platform. The virtual model exhibits a more regular shape, and during actual installation and testing, deformation is nearly imperceptible. Therefore, only primary interference inspection is needed to ensure accurate and error-free installation of the physical model, as illustrated in Figure 11.

**Figure 11.** Virtual assembly diagram of base frame of connector model testing system.

Upon the completion of the primary framework, the assembly of components such as the bow connector, stern connector, transverse reaction frame, and cylinder actuator demands a more refined method beyond simple initial interference checks. As shown in Figure 12a, taking the virtual assembly of the bow connector and stern connector as an example, the system prompts that they have collided without any contact. This is because it is difficult to fully fit the actual surface of the model into the OBB algorithm during the virtual assembly process. Therefore, the STL format file obtained by scanning the bow and stern connectors with a laser 3D scanner is imported into the Unity3D platform. The two models are attached by adjusting the 3D coordinates, and the Devillers and Guigue algorithm is used to determine whether there is interference between the two. Finally, the experimental system was assembled truthfully in the virtual space, providing a basis for accurately determining the dimensions of the pin axis in the future, as shown in Figure 12b.

The pin, which connects the bow and stern connectors, is a critical structural component of this experiment. To ensure the successful completion of the final assessment, it is imperative to employ advanced interference-checking algorithms. Based on the previous virtual assembly, commercial CAD software (25.0) was used to construct the initial pin shaft dimensions, and the system was imported to assemble with the bow and stern connectors. Advanced interference detection algorithms were used to quickly determine the pin shaft dimensions, with an assembly accuracy of up to 0.01 mm. In Figure 13a, the three-dimensional coordinates (in meters) of one of the virtual models are circled in red. By approaching the two models, an alarm will be triggered when interference occurs, and the distance between the two models is nearly 0.01 mm. This precision determines the optimal machining dimensions of the pin, thereby facilitating the installation of the entire testing system and ensuring the smooth execution of the experiment.

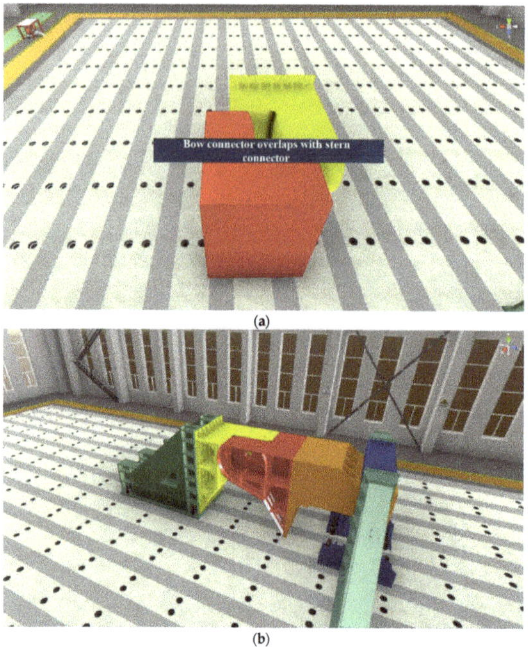


**Figure 12.** Virtual assembly diagram of connector model test system. (**a**) Bow connector and stern connector interference misdetection; (**b**) intermediate interference-checking algorithm is applied to effectively detect collision.

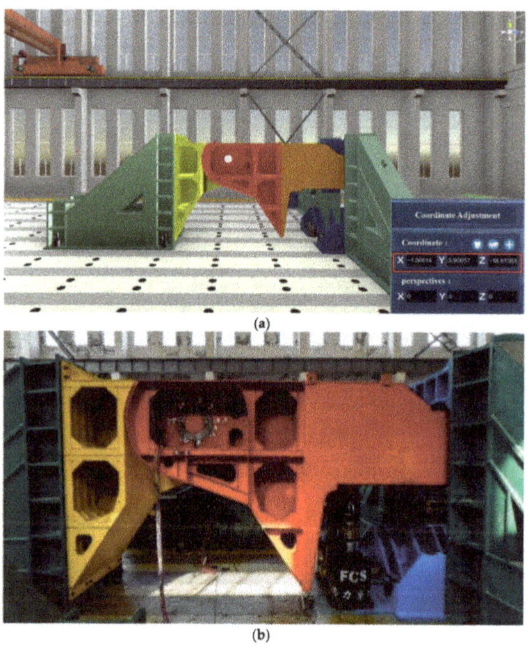


**Figure 13.** Comparison of physical and virtual test scenarios for connector model test system. (**a**) Virtual test scenario; (**b**) physical test scenario.

## 4. Conclusions

In response to the demand for the development of virtual integration test technology for ship structures, a high-precision virtual assembly method is developed. A three-level interference-checking strategy that incorporates a combination of the surrounded box, triangular intersection, and differential triangle facets algorithm is utilized. The efficacy of the proposed method is validated through a connector model test, the conclusions are as follows:

- The rough interference checking of the object is to be measured by the initial interference checking based on OBB-type enveloping boxes. This method is primarily suitable for conventional virtual models with regular shapes of enveloping boxes such as reaction frames and beams. While the algorithm demonstrates high computational efficiency, it is susceptible to misjudgment in complex structural models.
- The triangular face sheet detection is conducted as the intermediate interference checking based on the Devillers and Guigue triangle intersection algorithm for the models potentially interfered with after the initial interference check. This technique is suitable for providing a qualitative judgment of the interference status of all the models.
- Advanced interference checking relies on the differential triangle facets algorithm and builds upon intermediate interference checking. It quantitatively calculates the interference scale of key objects of interest, effectively guiding physical entity processing and manufacturing while ensuring smooth test implementation.
- The virtual assembly method presented in this paper enables the precise virtual assembly of the physical testing systems within virtual testing environments, and the assembly accuracy can reach 0.01 mm. The feasibility of the virtual assembly method is verified, which contributes to decision making during the on-site installation of the physical experiment, enhancing operational efficiency and safety, and reducing experimental costs.

From the application case, it can be seen that this method can achieve the virtual assembly of physical testing systems in virtual testing scenarios. By simulating the virtual assembly process, the feasibility of the experimental assembly plan is fully verified, guiding the on-site installation decision making of physical experiments, greatly improving the operational efficiency and safety of on-site experiments, and reducing experimental costs. In the future, further research will be conducted on the applicability of this method on different complex marine engineering structures, and the efficiency improvement of this method will be quantitatively studied. At the same time, a real-time collision interference detection method based on measured data will be developed to achieve interference warning between the structural test model and supporting fixtures under large deformation conditions, and to assist in the further development of virtual assembly technology in the field of ship structural testing.

**Author Contributions:** Conceptualization, Z.J. and P.W.; methodology, Z.J.; software, Z.J.; validation, J.P.; investigation, J.P.; data curation, Q.Z.; writing—original draft preparation, Z.J.; writing—review and editing, Y.D.; visualization, J.P.; supervision, Z.J.; project administration, Z.J.; funding acquisition, Q.Z. All authors have read and agreed to the published version of the manuscript.

**Funding:** This was funded by the Project of Virtual–real fusion Testing Technology for Ship Structure Performance (WDZC70202030202).

**Institutional Review Board Statement:** Not applicable.

**Informed Consent Statement:** Not applicable.

**Data Availability Statement:** The data presented in this study are available on request from the corresponding author. The data are not publicly available due to privacy concerns.

**Acknowledgments:** We appreciate the anonymous reviewers and the associated editor for the constructive comments to improve the quality of the manuscript.

**Conflicts of Interest:** The authors declare no conflicts of interest.

## References


1. Sumi, Y.; Fujikubo, M.; Kawagoe, Y.; Kidogawa, M.; Kobayashi, K.; Nakajima, Y.; Nakano, T.; Sunouchi, M.; Tamura, K.; Ueda, N. *Interim Report of Committee on Large Container Ship Safety*; Committee on Large Container Ship Safety, Maritime Bureau of Japan's Ministry of Land, Infrastructure, Transport and Tourism (MLIT): Tokyo, Japan, 2013.
2. Wu, T.; Pei, Z.; Wu, W. Experimental Study on Progressive Collapse Characteristics of Hull Structure in Waves. *J. Wuhan Univ. Technol. (Transp. Sci. Eng. Ed.)* **2020**, *44*, 1062–1066.
3. Wei, P.; Li, C.; Jiang, Z.; Wang, D. Real-time simulation of ship structures based on virtual–real fusion interaction. *Ocean Eng.* **2024**, *295*, 116815. [CrossRef]
4. Boothroyd, G. Product design for manufacture and assembly. *Comput. Aided Des.* **2014**, *26*, 505–520. [CrossRef]
5. Wei, W. Research on the Key Technology of Virtual Assembly and Simulation Applications. Ph.D. Thesis, Dalian Maritime University, Dalian, China, 2013.
6. Sambhoos, K.; Koc, B.; Nagi, R. Extracting Assembly Mating Graphs for Assembly Variant Design. *J. Comput. Inf. Sci. Eng.* **2009**, *9*, 034501. [CrossRef]
7. Bonneville, F.; Perrard, C.; Henrioud, J.M. A genetic algorithm to generate and evaluate assembly plans. In Proceedings of the IEEE Symposium on Emerging Technologies & Factory Automation, Paris, France, 10–13 October 1995. [CrossRef]
8. Failli, F.; Dini, G. Ant colony systems in assembly planning. In Proceedings of the 2ndCIRP Intelligent Computation in Manufacturing Engineering, New York, NY, USA, 21–23 June 2000; pp. 227–232.
9. Park, J.; Huh, K. Optimal Task Sequence Planning for High Speed Robotic Assembly Using Simulated Annealing. *JSME Int. J. Ser. C* **2008**, *43*, 222–229. [CrossRef]
10. Siméon, T.; Laumond, J.P.; Nissoux, C. Visibility-based probabilistic roadmaps for motion planning. *Adv. Robot.* **2000**, *14*, 477–494. [CrossRef]
11. Wang, F.; Qiu, Y. A modified particle swarm optimizer with roulette selection operator. In Proceedings of the 2005 International Conference on Natural Language Processing and Knowledge Engineering, Wuhan, China, 30 October–1 November 2005. [CrossRef]
12. Yeh, H.Y.; Thomas, S.; Eppstein, D.; Amato, N.M. UOBPRM: A uniformly distributed obstacle-based PRM. In Proceedings of the 2012 IEEE/RSJ International Conference on Intelligent Robots and Systems, Vilamoura-Algarve, Portugal, 20 December 2012. [CrossRef]
13. Zhang, R.; Lv, Q.; Li, J.; Bao, J.; Liu, T.; Liu, S. A reinforcement learning method for human-robot collaboration in assembly tasks. *Robot. Comput. Integr. Manuf.* **2022**, *73*, 102227. [CrossRef]
14. Bilberg, A.; Malik, A.A. Digital twin driven human-robot collaborative assembly. *CIRP Ann.* **2019**, *68*, 499–502. [CrossRef]
15. Zhang, L.; Huang, X.; Kim, Y.J.; Manocha, D. D-Plan: Efficient Collision-Free Path Computation for Part Removal and Disassembly. *Comput. Aided Des. Appl.* **2013**, *5*, 774–786. [CrossRef]
16. Liu, X.; Weng, X.; Chen, H.; Cao, L. Accurate collision detection using improved octree algorithm. *J. Comput. Aided Des. Graph.* **2005**, *17*, 5. [CrossRef]
17. Lin, M.C.; Gottschalk, S. Collision Detection Between Geometric Models: A Survey. *Vis. Comput.* **1998**, *11*, 542–561. [CrossRef]
18. Jiménez, P.; Thomas, F.; Torras, C. 3D collision detection: A survey. *Pergamon* **2001**, *25*, 269–285. [CrossRef]
19. Chang, J.W.; Wang, W.; Kim, M.S. Efficient collision detection using a dual OBB-sphere bounding volume hierarchy. *Comput. Aided Des.* **2010**, *42*, 50–57. [CrossRef]
20. Palmer, I.J.; Grimsdale, R.L. Collision Detection for Animation using Sphere-Trees. *Comput. Graph. Forum* **1995**, *14*, 105–116. [CrossRef]
21. den Bergen, G.V. Efficient Collision Detection of Complex Deformable Models using AABB Trees. *J. Graph. Tools* **1997**, *2*, 1–13. [CrossRef]
22. Gottschalk, S.; Lin, M.C.; Manocha, D. OBBTree: A Hierarchical Structure for Rapid Interference Detection. *ACM SIGGRAPH Comput. Graph. Annu. Conf. Ser.* **1996**, *30*, 171–180. [CrossRef]
23. Zachmann, G. Rapid collision detection by dynamically aligned DOP-trees. In Proceedings of the IEEE 1998 Virtual Reality Annual International Symposium (Cat. No. 98CB36180), Atlanta, GA, USA, 14–18 March 1998. [CrossRef]
24. Wan, H.; Fan, Z.; Gao, S.; Peng, Q. A parallel collision detection algorithm based on hybrid bounding volume hierarchy. In Proceedings of the CAD/Graphics'2001 the Seventh International Conference on Computer Aided Design and Computer Graphics, Kunming, China, 22–24 August 2001.
25. Kim, D.; Heo, J.P.; Huh, J.; Kim, J.; Yoon, S.E. *HPCCD: Hybrid Parallel Continuous Collision Detection Using CPUs and GPUs*; Blackwell Publishing Ltd.: Oxford, UK, 2009. [CrossRef]
26. Pabst, S.; Koch, A.; Straer, W. Fast and Scalable CPU/GPU Collision Detection for Rigid and Deformable Surfaces. *Comput. Graph. Forum* **2010**, *29*, 1605–1612. [CrossRef]
27. He, Y.; Chen, Y.; Zou, X.; Zhou, D.; Liu, J.; Yan, D. A clearance control and interference repair method for ship rib plate assembly based on reverse engineering. *Ocean Eng.* **2024**, *309*, 118402. [CrossRef]
28. Lau, K. Visualising the front line of shipping's technology transformation. *Ship Offshore* **2024**, *4*, 28–29.

MDPI AG
Grosspeteranlage 5
4052 Basel
Switzerland
Tel.: +41 61 683 77 34

*Journal of Marine Science and Engineering* Editorial Office
E-mail: jmse@mdpi.com
www.mdpi.com/journal/jmse

www.ingramcontent.com/pod-product-compliance
Lightning Source LLC
Chambersburg PA
CBHW061603210926
51850CB00009B/2199
* 9 7 8 3 7 2 5 8 6 5 3 6 9 *

MDPI AG
Grosspeteranlage 5
4052 Basel
Switzerland
Tel.: +41 61 683 77 34

*Pharmaceuticals* Editorial Office
E-mail: pharmaceuticals@mdpi.com
www.mdpi.com/journal/pharmaceuticals